NEW PRODUCT DEVELOPMENT: DESIGN AND ANALYSIS

NEW DIMENSIONS IN ENGINEERING

Editor
RODNEY D. STEWART

SYSTEM ENGINEERING MANAGEMENT
Benjamin S. Blanchard

NONDESTRUCTIVE TESTING TECHNIQUES
Donald E. Bray

LOGISTICS ENGINEERING
Linda L. Green

NEW PRODUCT DEVELOPMENT: DESIGN AND ANALYSIS
Ronald Eugene Kmetovicz

INDEPENDENT VERIFICATION AND VALIDATION: A LIFE CYCLE
ENGINEERING PROCESS FOR QUALITY SOFTWARE
Robert O. Lewis

DESIGN TO COST
Jack V. Michaels
William P. Wood

OPTIMAL INVENTORY MODELLING OF SYSTEMS:
MULTI-ECHELON TECHNIQUES
Craig C. Sherbrooke

COST ESTIMATING, SECOND EDITION
Rodney D. Stewart

PROPOSAL PREPARATION, SECOND EDITION
Rodney D. Stewart
Ann L. Stewart

NEW PRODUCT DEVELOPMENT: DESIGN AND ANALYSIS

RONALD EUGENE KMETOVICZ

A Wiley-Interscience Publication

JOHN WILEY & SONS, INC.

New York • Chichester • Brisbane • Toronto • Singapore

Copyright © 1992 by John Wiley & Sons, Inc.

Library of Congress Cataloging in Publication Data:
Kmetovicz, Ronald E., 1947–
    New product development: design and analysis / Ronald E.
Kmetovicz.
        p.   cm.—(New dimensions in engineering)
    "A Wiley-Interscience publication."
    Includes bibliographical references and index.
    ISBN 0-471-55536-3 (cloth):
    1. Design, Industrial—Management. I. Title. II. Series.
TS171.4.K58   1992
658.5'752—dc20   91-28941
                                                  CIP

Printed and bound in the United States of America by Braun-Brumfield, Inc.

10 9 8 7 6 5 4 3 2 1

FOREWORD

New Product Development: Design and Analysis is an important and timely addition to the "New Dimensions in Engineering" series. Kmetovicz provides an organized, methodical, and systematic approach to bringing new products and services into the marketplace. In keeping with the intent of the Series, this book provides a disciplined, engineering type viewpoint of a subject that is vital in today's competitive economy. The author, having been the chief executive officer of a company that specializes in shortening time-to-market, knows how to implement this systematic approach. And he has articulately conveyed the methods and techniques in this volume. When combined with the other systems engineering and multidisciplinary technologies represented by the Series, the information contained in this book offers the reader a clear path to product/service timeliness and product/service excellence. For engineers, entrepreneurs, managers, and marketeers, this book provides essential insights and a clear road map to new product or service success. We commend it to you for in-depth study as well as for use as a continuing reference.

RODNEY D. STEWART

Series Editor
January 1992

v

PREFACE

In general, this book explains how all the pieces of the new product development puzzle fit together. The "black art" of product development is exposed—it now becomes a science. For the first time, a product development organization has a comprehensive, cross-functional "how to" contained within one book. Inexperienced organizations can apply the material "as is" and be successful developers of new products. Experienced organizations will find themselves using the text as a comparative reference to measure, analyze, and improve existing practices. Organizations that implement the suggestions within the text will become leaders in their markets.

The text assumes that the reader is part of a new product development team. It does not matter what function employs you or what your job is called. If you have some doubts about what you are supposed to do and wonder what others around you are supposed to do, this book brings it all together for team members and their managers. It has a cross-functional focus that details roles, responsibilities, and output from each member of the team as the development of the new product moves from idea to eventual delivery to the customer.

The new product development process described is based on system theory. It is designed to improve organization time to market performance, provide significant gains in the control of development costs, and help a new product development organization produce a product with performance appropriate for the intended application. This new product development process works vertically through an organization's management structure and horizontally across all functions that contribute to the realization of the new product.

From a global perspective, the book is organized around the New Product Development Process Matrix (NPDP Matrix). This matrix has the organization's functions on its horizontal axis and time on the vertical axis. The time axis is subdivided into definition, planning, and execution segments. The flow through the text follows the time axis of the NPDP Matrix. Initial chapters focus on definition. Once the product's definition has been established, discussion moves to planning and then execution. Over time, individuals and managers are given the insight about what they and their functional groups should be producing to ensure effective cross-functional, concurrent, development of the new product.

Supporting the global organization are a number of visual tools and models that make understanding this diverse and complex subject very easy. Concepts like the Product Classification Matrix, Definition Matrix, New Product Development Sheet (NPDS), Resource Allocator (RA), Plan Synthesis Tool Kit (PLASTK), and the KMET Chart are introduced to maintain reader motivation while making the entire subject easily comprehensible. All efforts have been made to simplify rather than complicate the material presented.

Top management will find Chapters 1, 2, and 3 to be of particular value as they move to make the changes required by their organization. Overall, this group can use the information distributed throughout the text to help shape and guide their organization.

Members of Strategic Business Units will find the material in Chapter 2 to be of organizational and strategic importance. Once read and comprehended, they will know the role they must perform within their organization.

A matrix format guides cross-functional participants through the entire product definition process. They are shown how to proceed and how to evaluate their effectiveness. By following suggested topics, they are certain to improve the quality of the product's definition so that it can be appropriately staffed and developed in a timely manner. Chapter 3 guides these individuals in establishing that the project team works on the "right things."

Planning receives a completely new treatment—the topic is taken into the twenty-first century. New techniques show how to make the planning process distributed, fun, and fast. No longer is planning reserved for a few individuals within the organization in possession of a few key skills. The techniques make it possible for planning to be done by just about everyone on a concurrent, cross-functional, basis. The planning process becomes shared and distributed—the characteristics that are needed to make it effective.

Once a plan is produced, planners are given the information, tools, and techniques they need to determine the quality of the plan they produced. The text provides a simple and yet rigorous method that ensures that plans are based on reality and reason rather than idealism and wishful thinking. Chapters 4 and 5 are dedicated to plan synthesis and plan analysis. These chapters should be read by all members of the organization in order to become familiar with how they contribute to plan synthesis and plan analysis.

The cross-functional teams that transform the product's definition into a marketable product are shown how their performance is measured and enhanced by a combination of simple tools and methods. Tracking group performance over time becomes a friendly and easy task as needed adjustments are inserted into the development process to maintain team effectiveness on a concurrent, cross-functional, basis. No longer is it necessary to second guess where problems reside and what should be done about them. Data about the execution phase are presented to give early warning and detection. Members of the new product development execution team will find topics in Chapters 6 and 7 to be of particular value.

Originators of ideas are shown how to describe and promote their ideas within the corporate environment. Suggestions are given that make it very clear how ideas should be presented and evaluated. Originators of ideas will find Chapter 2 to be of interest.

Once an idea has been selected for development, a project manager is assigned to lead the effort. This individual will be interested in all the material within the book, but will find Chapters 4, 5, 6, and 7 to be of greatest value. In these chapters, project managers are shown how to build their team and guide it along toward the successful completion of its mission.

Suggestions are made throughout the text on when decisions should be made and who should make them. No chapter is exempt from this topic. As such, key decision makers will have an active interest in the entire text to guide and improve their skills.

RONALD EUGENE KMETOVICZ

Cupertino, California
January 1992

CONTENTS

NEW PRODUCT DEVELOPMENT: DESIGN AND ANALYSIS

1
OVERVIEW

In this introductory chapter, you will learn about

The purpose of the text and its application focus.

The major variables affecting new product development:
Change and complexity
Change and conformity
Invisibility
Expense control
Need for speed
"Can Do"

The problems present in the new product development environment and the most common reasons for their occurrence.

Models of organizational structures.

The "matrix organization."

Four product development classifications and their impact on process timing relationships.

Paths to follow that *might* lead to process success.

Paths to follow that *will* lead to process success.

High-tech products that come to market six months late but on budget will earn 33% less profit over five years. In contrast, coming out on time and 50% over budget cuts profits only 4%.

McKinsey and Company

1.1 PURPOSE

The purpose of this book is to document and explain an effective new product development process that enables new product development teams to transform product ideas into product reality in a shorter time than was previously thought possible within their operational environment.

Additionally, the book supplies information that will assist new product development team participants execute their job responsibilities with enhanced precision in the delivery of output that is within budget and meets performance objectives.

To achieve the above two objectives, I've created a systematic new product development process that is firmly based on system theory. This process is designed to improve organizational time to market performance, provide significant gains in the control of development costs, and help a new product development team produce a product with performance appropriate for the intended application. The new product development process described in this text works vertically through an organization's management structure and horizontally across all functions that contribute to the realization of the new product. Some organizations may initially have difficulty in adopting the entire process. In this case, localized adjustments can be produced to tailor the process to the task at hand. The material to follow has been applied on real product development projects over a period of years and a variety of conditions to test its fundamental suitability, function, and accuracy. Care has been taken to address the needs of a diverse product development audience. Software, hardware, and system product development will benefit from the material contained in the text. The modeling and visualization techniques utilized should prove to facilitate understanding and application.

For any system to be analyzed, characterized, and understood, a model is generally required. In the case of this text, the system is modeled using a matrix and network technique. "Matrix" management has been actively studied and documented by industrial and academic groups. The two-dimensional matrix representation has many strengths. A primary one is its ability to record responses at the intersections of rows and columns. A limitation of the "matrix" approach is its inability to represent information flow as a function of time. To overcome this limitation, I use network theory as a supporting system modeling technique. Thus, the marriage of the matrix and the network allowed me to produce a visual system model that can be applied to meet the two primary objectives of this text, identified in the first paragraphs. Moreover, the matrix and network models provide a bridge to effective tool development, utilization, and support.

Tools play a significant role in the application and administration of any systematic process. This process is no exception. It would be preferable to have this process fully computerized; but, for now, it does not appear to be achievable. As you read on, you will discover that the new product development process requires a great amount of intelligence that is not within the state of the art of present computer technology. To overcome this limitation, the system was tuned to function with readily available computer systems that require manual support. When human interaction was needed, all attempts were made to keep the work that needed to be done visual, entertaining, and stimulating.

1.2 VARIABLES AFFECTING NEW PRODUCT DEVELOPMENT

For a systematic process to be effective, it must be adaptable to the forces that keep it in operation. In the past few decades and, particularly, the decade of the 1980s, the process of creating new products has undergone dramatic transformations. New driving forces are being created and existing ones are being modified in magnitude and phase. Those of us involved in this process know that it is far too easy to miss schedules, blow a budget, or deliver a flawed product. We've been told about the trade-off between performance, time, and money, and that the precise specification of two will intrinsically define the third. As we've moved forward on to new product development assignments, the truth contained in the previous statement revealed itself in many mysterious ways. It is our experiences that have taught us the true nature of the new product development process, experiences that many of us have little hope of documenting and sharing. Thus, as we gain experience in the field, we tend to internalize the knowledge we gain. I call the inward nature, the true substance, and the constitution of the new product development process its essence.

I've had the opportunity to participate within this process for 20 years. Over this period, I have observed that the new product development process

1. *Is* in place to create change.
2. *Is* complex and getting more complex.
3. *Is* attempting to conform to outside pressures.
4. *Is*, in many ways, invisible.
5. *Is* very expensive.
6. *Is* constantly under pressure to make it move faster.
7. *Is* reserved for people with a "can do" attitude.

This list of seven items is the essence of the new product development process. With the exception of discussing change directly, I will provide some information on the other six items to assist you in understanding the foundation of the process.

1.2.1   Change and Complexity

The creation of new products is not easy, and most likely will never be. The primary factor contributing to the difficulty of new product development is managing change; and yet the management of change is a crucial part of the new product development methodology. Those with a strong engineering background immediately associate change with advances in technology. While technology is one part of the change mechanism, it must be recognized that it is not the only element.

New product development is a "world" awareness adventure. Significant new product development efforts are usually no longer performed in the workshop, garage, or kitchen. The days of Franklin, Wright, Edison, and the like seem to be behind us. Sure, individuals still make key discoveries, but converting individual science/research efforts into new products is now a team endeavor that has global implications. Those of us involved in the creation of new products are now required to scan the globe for new insights. We must look at markets, customers, distribution channels, marketing techniques, finances, government regulations, forming standards, following standards, and so on. Furthermore, our vision must be kept focused on technological developments. Clearly, we must operate in an environment that is undergoing constant change while increasingly growing in complexity. I'll work with two examples, recorded music and the personal computer, to help explain my thoughts on change and complexity.

Looking at recorded music helps illustrate the growth in complexity that change has produced from a new product development perspective. The industry's evolution can be tracked from Edison's wax cylinder to the present-day compact disk player. Between the two extremes is the phonograph. It started in the 1920s with the 78-rpm record that offered up to five minutes of play time on a side. Thirty years later, the 33-rpm record and high fidelity appeared on the market. Within 10 years, hi-fi was replaced by stereo recordings. Analyzing the design and construction of all electromechanical players over this evolutionary period is a relatively straightforward process. They basically consisted of a motor, a platter to hold the recorded media, an electromechanical magnetic transducer, a mechanism to guide the transducer, and a few simple controls. It was possible to look at the device and begin to understand its functionality. In general, the technology could be explained to the average technical college graduate who, in turn, would feel comfortable with the explanation. Subtle complexity could be added to make performance improvements. Over the years we were witness to many incremental improvements in this device. The high end in performance was understood by few and fully appreciated by a narrow niche in the marketplace—the audiophile. The best performance available from this analog technology gave us 65 dB of dynamic range, 30 dB of channel separation, linear tracking, and the ability to listen to dust, wear particles, and contamination. Sixty years after its humble beginnings, the technology was forced into decline by the next generation replacement, the CD. However, it left behind a complex infrastructure that could deliver recorded material in the CD format with little change or alteration.

The infrastructure consisted of entertainers, recording studios, prerecorded music libraries, media manufacturers, promotion agencies, wholesalers, retailers, mail order outlets, and other "total solution" products and services needed by this market. The infrastructure has some technological dependencies, but, by and large, it functions mostly detached from the record, manufacture, and playback technology of analog or digital. Therefore, the new technology was adapted to work within the existing system to help guarantee its market success by keeping nontechnological change to a minimum.

When the compact disk player arrived on the market, it was easy to promote its benefits of greatly improved sound quality, extended playing time, reduced media size, no media wear, added functionality, and an ever-expanding selection of media from worldwide sources. Technical specifications were published, but hearing 96 dB of dynamic range without distortion was generally enough to convince many in the market to switch to the new format. The compact disk became a market success and its technological evolution continues.

From a detailed technological perspective, fully understanding all of the functionality within a compact disk player is out of reach for the average technical college graduate. The compact disk player does have a motor and some have simple controls. Any analogy with the phonograph ends here. The new player consists of lasers, microcontrollers, servo mechanisms, digital integrated circuits, internal program codes, digital signal processing integrated circuits, digital-to-analog converters, sampling circuits, and linear amplifiers. The careful systemization of these various forms of technology produced the compact disk player as we know it today. For consumers, this detail is hidden. For technical managers, the detail presents itself as an educational and management challenge! Significant change has arrived and it is complex. A total detailed understanding of all aspects of this system is probably beyond most technology managers. Understanding can generally be achieved to a given level when this system is hierarchically segmented. Some individuals will comprehend the overall function of the system. Others will understand, in detail, the functionality of subsets of the system. It's not likely that one individual will understand *the system* and the full detail of *all the subsystems*.

This example illustrates how technological change and complexity can be managed while minimizing change in nontechnology related areas necessary for product success. It would be possible to go back to the early 1900s to learn how it all began. Instead, we will examine a more contemporary example to explore the creation of a new market with new technology. To do so, contrast the personal computer to that of the compact disk to begin to visualize the added complexity needed to make the personal computer a successful product in the market. The developers of the early personal computers had to develop the technology, as well as the entire infrastructure, to deliver the product to the consumer. Try and recall the makeup of the personal computer industry in the late 1970s.

You may remember in 1979 (the microprocessor having appeared years earlier) hearing about this new firm in Cupertino, California, called Apple. We were told they were making a personal computer. At that time, I was busy doing my computing on a mainframe or on a desktop workstation. We were programming microprocessors using assembly language to use them in instrument control applications. General-purpose application software was almost a foreign concept in research and development. Most of the time I had to solve my own problems by developing models and then writing FORTRAN or Basic programs to find solutions. I wasn't alone. "Real" computer users wrote their own software. Meanwhile the Apple II appeared on the market. It did not have the computing power to meet my needs. Hewlett-Packard saw to it that I had access to far better equipment. I viewed the Apple II as a toy. While I was not impressed, Apple pressed on to:

1. Develop application software.
2. Educate the marketplace on simple things the machine could do.
3. Create a marketing and distribution system.
4. Build a customer base in the education and home environment.
5. Convince onlookers that there was a market in personal computers.
6. Create and develop an entirely new market.

Apple, at that time, was working to define their business, their products, and their market. I can speculate that change and complexity were constant friends and foes. This early work was being performed in an environment where nothing was stable and where nothing was simple. Given such circumstances, one would have expected that a number of mistakes would be made along the way and that the development process would have to be flexible to accommodate numerous redirections.

We saw the rise of the Apple II and the distribution network established by Apple. Application software at the time was limited to simple word processing, games, and some basic business tools. In 1981, the CPM operating system appeared along with a number of new personal computer manufacturers that bet their companies on the success of CPM. These companies were not with us very long! IBM and Microsoft appeared in 1982 with a blockbuster new product, the PC with MS-DOS. Following the appearance of this machine, we saw the development of the PC software industry. With the arrival of the spreadsheet and better word processing, the personal computer market made a shift from home and education to business. Businesses, large and small, found that they needed the new capability. The market expanded dramatically. Coupled with this expansion was the creation of a software retail distribution channel. Existing business retailers made attempts to meet consumer demands for hardware and software. For the most part, the existing retail channels did not function properly. The consumer, and business, chose to do business with a new retailer that focused on sales, service, and support of PC hardware and software. We were witness to the rise of Businessland, Computerland, and the like. New hardware, software, and system products are appearing at an ever-increasing rate. This industry and market are just beginning.

Prices on base hardware configurations decline while high-performance machines command high price premiums. The software industry has grown to meet consumer demand for improved and new applications. Hardware components that make up these computers are topics of international trade disputes. Individual computers have given rise to networked systems. The machines get smaller and faster. Market growth continues with no end in sight. Reflecting on the developments that have transpired over the past 10 years is a source of constant amazement! The personal computer industry certainly appears to be changing rapidly and growing more complex from a product development perspective, while the infrastructure seems to be stabilizing and rooting itself on a global basis.

Any system that deals with the development of new products must be capable of effectively dealing with change and complexity at either of the extremes that have been described. I've elected to include what I called the "infrastructure" in the matrix and network model. In applications where the infrastructure is in place and functional (compact disk example), the matrix can serve as a guide to performing the needed work to satisfy requirements. For organizations/operations that must start from the beginning, the matrix can be used as a model for the design and development of an efficient nucleus for a new product development system. Infrastructure displays itself in the matrix by including marketing, manufacturing, sales, and so forth, along with the more traditional technology areas of systems, hardware, and software. The new product development system described in this book could have been used effectively on the compact disk or the personal computer with only minor modification.

1.2.2   Change and Conformity

Conformity is polarized. Let's call the first "negative conformity"; its opposite is termed "positive conformity." In either case, the new product development process is forced to conform to constraints that were developed by different people over a period of time. Because the rules and regulations are man-made, they are subject to analysis, challenge, and criticism before being accepted and applied. Generally, positive conformity is more readily accepted in the product development environment than the negative type. Let's first explore a few of the facets associated with negative conformity.

Negative conformity typically starts with a directive from someone. The individual, or group, can be a customer, institution, bureaucracy, regulatory agency, corporate officer, manager, or co-worker. It can be forced upon the new product development process without rhyme or reason. Because of the large number of input sources into the process, the probability of being required to conform to some "off the wall" requirement is quite high. I'll share with you some of my most common frustrations. By doing so, it is hoped that I can encourage you to identify those that are common in your work environment.

At the top of the list is the "that's the way we do things here" syndrome. You'll find this characteristic in many mature organizations that have established pieces of a new product development process. Usually, the process has evolved over a period of time. Many of the individual segments of the process were learned by making major mistakes. As each error was detected, the process was adjusted to prevent the error's occurrence in future new product development programs. With each subsequent error, the process has another insurance layer added as it grows in complexity. It eventually gets to the point where it will guarantee that no errors will be produced by completely eliminating the possibility of producing any output from the system! The best offensive is to determine the reasons for all the added layers of process insurance. When the justification makes sense and it's found to be an overall process accelerator, leave it alone. When it does not, it is a signal to prepare a logical, well-reasoned alternative and attempt to make a revision in the process. It is extremely important to stay proactive on these issues. As you begin to make changes and establish a positive track record, it will become progressively easier to make future improvements. The proactive attitude will allow you to develop a trust between yourself and your employer. This trust makes it possible to make needed process improvements and alterations a reality.

Second is the "customer is always right" point of view. When dealing directly with the customer, the conformity strategy is relatively simple. Present your point of view as best you can; if it is accepted, great! If not, do it the way the customer wants it. Where it gets more difficult is when working with secondhand information. Here conformity is best reached by consensus. When dealing with ambiguity, the consensus decision process that collects input from a number of different sources produces amazingly effective results. A consensus decision to conform in a given way is usually the best answer. There are times, however, when agreement cannot be reached using a consensual procedure. If this happens, don't let it drag out. It's much more important to move forward once everyone has had the opportunity to speak their mind than it is just drift in a sea of indecision. A prompt decision made by a person with necessary organizational power is usually best for the overall new product effort.

Third, there is a tendency in business to aim for what seems to be the "safe bet." This form of negative conformity will sometimes manifest itself in the creation of the new product development committee. To quote Dr. Porsche: "Committees are, by nature, timid. They are based on the premise of safety in numbers; content to survive inconspicuously, rather than take risks and move independently ahead. Without independence, without the freedom for new ideas to be tried, to fail, and to ultimately succeed, the world will not move ahead, but live in fear of its own potential." The new product development process requires people that can see beyond risk, manage uncertainty, and motivate others to shrug off negative conformity. I have seen few successful new product development programs where certain individuals did not assume a great deal of personal risk to help make the output of the process a success. Process participants need to feel the challenge and the victory! Doing "safe" things does not elevate the inner spirits of people that enjoy working within the new product development environment.

The last major conformity issue of this type is to adapt to "It's required by the management information system (MIS)." Notice the use of the word *adapt*! There may be little you can do to change the requirement, but there may be a lot you can do with the quality of the information that you deliver to the system. This positive attitude will allow you to create systems and techniques that produce quality information for the MIS (incidentally, I found that high quality significantly reduced the quantity that was required). It is very important to focus on information-generation issues rather than computer system issues. There are far fewer people that are capable of creating accurate information on the new product development process than there are those who can operate information systems.

The preceding four examples of negative conformity each contain suggestions on the actions you can take to convert a negative situation to a more positive form. This is done to indicate that efforts should be initiated to eliminate this type of contamination from the new product development process. You will find no provision in the process contained in this text to deal with this negative form, but you are encouraged to take immediate action to make an effective transformation to the positive form by being proactive and creative.

"Positive conformity" can be an asset to the new product development process. Developing a business around real customer needs is a form of positive conformity. Working to establish world standards is another form. Think of the benefits derived from standards like MS-DOS, UNIX, and X.25, to mention only a few. Designing a product to follow world standards can be an excellent step in helping advance standards, reduce development cost/risk, and improve marketability. Manufacturing products with standard components and standard processes may assist in design, manufacturing, and support aspects of the new product development process. Using standard tools may be the answer to nagging service problems. Positive conformity can be the correct solution to many new product development process problems.

This text provides a documented new product development process for readers to follow. While not written as an absolute standard, it can help you establish a standard within your product development environment. Coverage of essential elements of the new product development process is provided. Your organization can use it as a reference to aid in creating your first new product or it can be used as a point of comparison to make alterations to your existing process. In either case, it will help you solidify certain aspects of the new product development process into a standard within your organization. Having a standard product development process will be of extreme competitive value!

1.2.3 Invisibility

For something to be invisible, it is usually thought not to be perceptible by the eye or easily discernible by the mind. Situations that can be modeled in two or three dimensions tend to be more visible than those that cannot meet this modeling criterion. Likewise, phenomena that can be mathematically modeled also tend to be more visible than those that cannot be described by a system of closed equations. The new product development process embraces many concepts that are basically invisible to the development team. How does one go about modeling human characteristics such as ego, optimism, intelligence, creativity, motivation, and self-discipline? Where do we begin to make visible such organizational qualities as communication, teamwork, leadership, force multipliers, roles within the work society, and structural interactions? What about the invisible structure associated with technology, markets, economics, and systems that is virtually impossible to model in the development of new products? At times, I find myself searching for quality answers to the questions just raised. I have yet to find anything concrete to provide definitive answers. There are, however, techniques, discussed below, that keep the invisible nature of product development in proper perspective.

My initiation into the direct management of invisibility came with my need to develop *systems* that contained significant amounts of *software*. I discovered that software was basically invisible from a modeling perspective. Geometric abstractions were sometimes very difficult, or impossible, to produce. Scale drawings, alternative form abstractions, system block diagrams, and the like could not be readily produced. I discovered that the reality of software did not lend itself to spatial 2-D/3-D modeling. It was not modelable like a discrete component analog circuit or a highly repetitive digital integrated circuit. Circuits had the property of being relatively easy to represent on paper or within computers. Generally, two or three views were adequate representations. I found myself, and my work group, attempting to model software using a variety of data flow, pattern dependency, timing, abstraction, and name–space techniques. We attempted to find links and transformations. We attempted to simplify our models. Gaining conceptual control seemed to be out of reach for all but the most simple of problems. We had difficulty modeling our own work for our own internal use. The problems of communicating our intentions to others who needed to know were exacerbated exponentially!

Solutions to the essence of software invisibility began to appear when we redis-

covered the practice of creative, rapid prototyping. Once we had produced prototypes, we were able to demonstrate them and measure them. The prototypes gave us
the ability to learn about our creations. We discovered strengths and weaknesses
about our design. Communication channels within the product development environment were opened (we actually produced things that we and others could see and
use). From the knowledge we gained, it became possible to model key aspects of
the software and its effect on system performance. Our modeling process became
more focused and efficient. Prototyping gave us the synergy we needed to break
through many conceptual roadblocks that stood firm when attacked only by analytic
modeling techniques.

Prototyping is a highly creative modeling process that produces visual and measurable results. Prototypes complement the visual nature of new product development. They serve their useful purpose through the early phases of product creation
and then tend to become less important as the real product begins to materialize.
For this reason, the matrix and network models place strong emphasis on prototypes during the definition phase. The model goes beyond the conventional techniques and suggests that prototypes of all aspects of the product be produced to aid
visualization of the total offering to the customer, support products, and process
that will be used in its design, development, manufacturing, sales, distribution,
and support. Any work that improves a development team's ability to visualize
improves the overall process. As product development moves from definition to
planning and execution, visualization techniques for these respective phases are supplied.

A planning technique will be described that actually makes the process of creating a top-down structured plan fun! While the project/program planners are having
fun making the plan, the plan materializes in a fashion that causes the creation process to become a visualization and communication vehicle. Once the plan is jointly
created by key members of the new product development team, it is then necessary
to measure the quality of the planning effort. This is done by a comparative process
with estimates from the definition phase and by making direct measurements on the
plan itself. The comparisons and measurements that are made on the plan are reported in a visual and intuitive format to facilitate clear understanding. Process visualization continues to grow and develop within the execution phase by tracking
and creatively reporting on the progress achieved and problems encountered by the
new product development team. Making "real-time" adjustments becomes a shared
visualization experience for the new product team. The tools and techniques that
make this a reality are fully described and explained within this text.

This book places strong emphasis on visual techniques to help you deal with the
invisible properties of the new product development process. In the chapters ahead,
matrix and network concepts will be introduced that will help you visualize the
entire process. The matrix and network models can be observed and measured using
a number of creative techniques. At each point in the process, it is possible to visualize the entire problem to assist in making those difficult new product development
decisions. The new product development process transforms itself from one that is

not observable and is difficult to control to one that is modeled, measurable, and in control.

1.2.4   Expense Control

The creation of new high technology products is a very expensive process! The average cost to support a single engineer for a period of one month ranges from 1.7 to 4.0 times the average engineering salary of the organization. Base-pay engineering averages are currently at $4000 per month in "Silicon Valley," California, which translates into loaded costs extending from $6000 to $16,000 per engineering person-month in this particular geographic region. This broad range of costs per engineering person-month is made evident to me each year when students in my research and development management class determine this figure for their respective companies.

Average person-month costs are at 2.7 times base pay. While base pay moves with time, the 2.7 scaling factor has remained basically constant over a three-year observation window. Companies that operate with costs significantly below the average tend to be young (less than five years in business) and relatively small in gross sales (less than $20 million). Much older companies and those that provide products and services to government and defense customers tend to populate the high end of the range. At the group average, salary contributes about 35 percent of the cost per engineering person-month. The remaining 65 percent can be structured into the categories of new product process overhead allocated to engineering, direct project expenses through the definition and planning phases, and direct project expenses through the execution phase.

Typical costs that new product development teams incur for payroll taxes and benefits, facilities, travel, recruiting, education reimbursement, office supplies, software, and support services are usually considered to be a part of the overhead structure of the project. Payroll taxes and benefits can run up to 20 percent of base salary. Travel is expensive; airline ticket costs, lodging, meals, and surface transportation can consume a significant amount of cash. Recruiting new talent into the organization is accompanied with a relatively steep price tag. At the college level, it is not unusual to interview 30+ students on campus to produce candidates for five in plant interviews. If things work out well, the five interviews will produce two permanent hires. For professionals, a 20 percent commission (based on the first year's salary) to an agency is the average rate. Office supplies are no longer limited to a few pads of paper and a few writing instruments. A word processor, laser printer, duplicating machine, toner cartridges, and 60# paper are likely to consume a considerable amount of money. Of course, consideration will be given to utilizing the latest style office environments and furniture. The more technically advanced organization may require workstations running design and process automation software. Once the cost of automation hardware is taken care of, the substantial flow of cash continues, this time for acquiring and expanding the utilization of various software tools. Most likely there will be a need to network computers and provide for electronic information transfer. Naturally, a few people will need to be available to keep the systems running and help provide the assistance that users require. As a rule of thumb, the overhead structure that I've just described doubles the average base pay of a new product development engineer—a $4000/month engineer now costs the organization $8000 per month.

To the base pay and overhead, it is still necessary to add direct project expenses through the definition, planning, and execution phases. In hardware development, engineering samples need to be built. Metal will be milled, folded, and turned. Printed circuit boards will be fabricated, loaded, and tested. Integrated circuits will be processed and fabricated. Software will be written. These building blocks of metal, silicon, electronics, and software are merged to form subsystems which are connected to create the complete system. Expenses associated with the creation of an engineering model are high and yet it is just the beginning of rapid cash flow. Once built, the system will be tested to verify that it performs to specification. Quite likely, it will not! Rework and retest will be required. Time and the associated money will continue to be consumed. As the design stabilizes, electronic and mechanical tooling will begin as others become active participants in the creation process. Quite likely, the new product development team will consist of players from research and development (hardware, software, and systems), manufacturing (scheduling, purchasing, fabrication, assembly, and test), marketing (product, sales, service, and documentation), and support (finance, personnel, and quality assurance). Each group will contribute to the rapid expenditure of funds. It is not unusual to have the direct project expenses just outlined run from 0.5 to 2.0 times base pay on average, depending on system complexity. The average person-month cost for an engineer earning a base pay of $4000 has now grown into the range of $10,000 to $16,000 per month.

At a $12,000/person-month ($144,000/person-year, $1.44 million/year for a 10-member team) expenditure rate, it is not difficult to understand why it is desirable to have a new product development process that can assist in the prediction and control of expenses. The process must focus on getting the product definition "right" so that the end objectives are made as clear as possible to minimize confusion, change, and rework. At the same time, it must create an ordered list of executable tasks that is "right" so that it runs near optimal efficiency. The process should guide you to work on the "right things" and to do "things right!" Doing both well keeps costs in control.

The best method I've discovered to keep costs in control and move projects along at an accelerated pace revealed itself when I analyzed staffing profiles of an assortment of projects. A historical staffing profile is the reconstruction of the people assigned to the realization of the new product over its definition, plan, and execute cycle. In most cases, the data can be gathered from old financial data and by tapping the memory of those still available within an organization.

Timely projects and those that had expenses well in control were characterized by low investment definition phases followed by a rapid escalation to full staffing to execute the project completely and quickly. To this general model of success,

I added a brief, but intense, planning phase. The resultant profile is shown in Figure 1.1.

This "ideal" staffing profile is characterized by a definition phase whose staffing magnitude is less than 20 percent of the full staff level and takes roughly 50 percent of the project's duration. The shape of the curve closely approximates a step function. It is important to note the inclusion of a short-time duration, intense, planning phase between the definition and execution phases.

As might be expected, the individuals assigned to do definition phase work will come from a relatively select pool of talent. These people will have to process customer, competitive, internal, and external information into a product description. This description must contain data about the performance, features, cost, price, and time to market of the product. Additionally, the interface description, specifications, technical risks, and investment details must be investigated and established. Most often, a cross-functional team of people with various skills in R&D, marketing, and manufacturing needs to be assembled to complete the definition phase. As elapsed time nears the 50-percent point, the cross-functional team establishes "what" is to be produced.

Once finished with this activity, the team can objectively view and utilize the product's definition as it enters a short, but intense, detailed shared planning period. During this period, all aspects of product realization are carefully studied and coupled to one another. Cross-functional planning teams determine work assignments down to the individual level. They identify the information flow between functions and establish key program milestones. As their work progresses, a network model of "how" the product will be created materializes. This planning phase is very different from traditional ways of planning. The methods are different because they have to produce better results faster than the old-style techniques. Plans

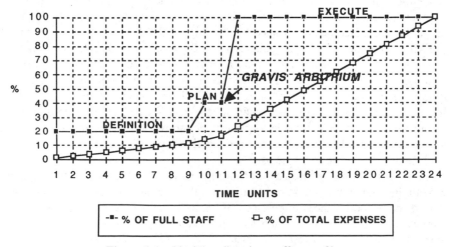

Figure 1.1   Ideal "step" project staffing profile.

must be done in "real time" and they must function in "real time." Future chapters will detail how these results are obtained.

The stage has been set for doing the "right things" and for doing "things right." With less than 20 percent of the total project funds expended, the project reaches a critical management decision point. The project team and the management team can now take time to answer the following critical questions: Is the definition of sufficient quality to justify the expenditure of the remaining 80 percent of funds? Does analysis of the plan show that it is based on reality? Will the organization commit the resources to compete the job within the remaining 50-percent time window? Should the effort go forward or should it be terminated? These are the questions that must be answered at the *gravis arbitrium* (important decision) point in time.

As you read on, you will find that considerable attention was paid to documenting a new product development process that is sensitive to getting the definition established early in time so that resulting plans and the execution of the plans are likely to undergo few revisions. At the same time, the process must be tolerant of change and be capable of dynamic adjustment. To do all of the above, the process places emphasis on being able to observe and measure the expenditure of funds in a friendly and constructive fashion to help guide the execution of the project. This "light" accounting approach tends to be favored by team participants and can produce results that will be within a few percentage points of actual (actual data usually arrives a few weeks to late). Absolute accuracy is sacrificed to give the process decision makers data in "real time" that can be used to extreme advantage. The process documented in this book is designed to eliminate, or significantly reduce, cost-containment problems.

1.2.5 Need for Speed

Response as a function of time is an important parameter to control in any process. The new product development process is no exception. New products are the foundation for continued business success in the high-technology field. Time to market (TTM) is the measure of an organization's ability to convert ideas into commercial product reality. It is a competitive requirement; it is a strategy; and, it is an organizational capability. Chris Meyer, with Strategic Alignment Group, defines effective time to market performance as "an ongoing ability to identify, satisfy, and be paid for meeting customer needs faster than anyone else." An effective new product development process must provide for timely idea-to-reality transformations. It must have a time-to-market focus to assist product developers in making the correct tactical, logistical, and strategic decisions.

Being an employee of Hewlett-Packard for 18 years and charting the company's growth due to new product introductions over that period of time has certainly convinced me of the business importance of regularly delivering new products to the marketplace. In the 1988 *Hewlett-Packard Annual Report,* it is stated that "more than half of 1988's orders were for products introduced in the past three years." If HP were to have had an across-the-board three-year setback in TTM (a highly unlikely scenario), the company would be far below its present revenue base. This would have been a business disaster! HP is certainly not alone in realizing that future business success depends on a steady and timely flow of new products to the marketplace. Review of just about every successful company's annual report will reveal their recognition of and dependence on a continuous supply of new products. From an organizational capability perspective, new products do not just happen; a system must be established for their creation. The system fuses products, processes, and people into a finely tuned unit to effectively satisfy customer requirements.

It's a complex system that rests on top of a research and idea base. People with differing interests and perspectives need to interact and quickly make difficult decisions based on rapidly changing information. Developing one alternative into "ultimate product" status is no longer an option. Successful decision makers have a number of alternatives under consideration simultaneously. Frequently, experienced advisors work directly with decision makers to expedite the decision-making process. When possible, the consensus decision-making style is employed, but if this style slows down the process, it is intelligently discarded in favor of a more rapid style to resolve time-perishable deadlocks. Furthermore, the players in the system know that no one action can be taken without it having an impact on something else. As a result, a great deal of effort is expended to integrate decisions made on one new product development effort with those from other efforts.

The time-sensitive new product development system exists to quench the thirst of customers who expect a competitive TTM. It should come as no surprise that business success correlates strongly with enterprises that deliver what the customer wants and needs at the time they want and need it. When a business can convert a customer requirement into a deliverable product or service within the customer's TTM frame of reference, the probability of securing the business is very high. As the organization's TTM extends, the odds of capturing the business diminishes. The point where TTM extension has a significant impact on business will be set by market conditions and the competitive environment. Clearly, if the competition provides a solution before an organization can deliver a product, the organization's TTM must be reduced on the next idea-to-product cycle for the business to remain in operation.

A piecewise linear model can be constructed that illustrates the concept by plotting the probability of product success as a function of TTM. Figure 1.2 shows that, for a certain TTM range, a high probability of success is retained. This TTM region will vary from business to business and from one product type to another. When concepts are new and not available in the marketplace, this region can extend for years. It is not unusual to have some excellent product ideas take an extended period of time to research and develop just to establish a product definition. The risk of extended exposure to time must be offset by the potential rewards that market success can bring. On more mature product developments, this segment of time can be rather short before TTM enters the competitive region. Once in the competitive region, the slope changes and business success becomes very sensitive to TTM. As the probability of business success declines, yet another steep change in slope is produced. More detail on the curve is presented below to aid in its understanding.

The thought process involved in constructing the shape of the curve was based

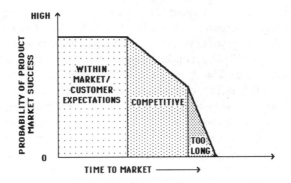

Figure 1.2   The relationship between product success and time to market.

on firsthand data I had available on the frequency synthesizer market and from outside sources. This data clearly show that there are four major product development classifications to consider. The classifications are created by using an *x*-axis field called "product concept" and a *y*-axis field called "market." Product concept is further divided into "familiar" and "new," whereas the market axis is segmented into "established" and "new." The resulting product classification matrix is shown in Figure 1.3.

The first cell to consider, formed by the intersection of the "new" row with the "new" column, is the creation of a "first-of-a-kind" product. In frequency synthesis, these were products that made major price/performance enhancements over previous product alternatives and provided completely new capabilities. They surfaced in the late 1960s in the radio frequency range and in the mid-1970s in the microwave frequency range. Their contributions were so significant that they changed the rules of business in existing markets and created many new opportunities. Other familiar examples to consider are products like the VCR, Teflon, Post-its, and microprocessors. The transformation from idea to product was measurable in terms of significant fractions of decades. It is not unusual to see efforts of this type take from five to seven years. Some developments like the television (10 years) and the telephone (15 years) take even longer. On developments of this type, the flat region of the curve shown in Figure 1.2 extends for a long period of time. The break into the competitive region will occur only if competition appears and the companies with a stranglehold on the market allow their position to erode.

In general, the people working on first-of-a-kind ideas tend to have a good awareness of where the competition is positioned in research and development. Individuals in a leadership position are usually quite good at predicting their position relative to others. This knowledge can be placed to good use in the establishment of TTM objectives. Likewise, a development group, in the catch-up mode, can use the knowledge available on their competitors to establish an appropriate TTM goal that is consistent with other business factors. Knowledge of the competition can be used to synthesize "what if" scenarios to determine the impact on product success as competitors' products appear in the market ahead of your introduction schedule. It

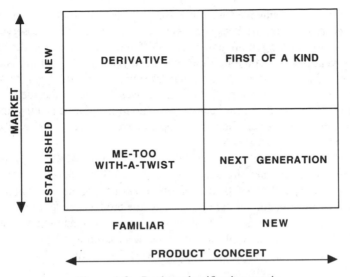

Figure 1.3   Product classification matrix.

is frequently possible to converge on the slope in the competitive region that provides some degree of analytic business assurance. More often than not, being best in this area is far better than being first with a product that doesn't work or doesn't make much sense.

It is usually difficult to involve the customer in first-of-a-kind programs. The challenge is to deliver a product that minimizes development time so that market returns can be generated as soon as is practical from the customer's perspective. Ideally, first-of-a-kind efforts should err on the side of doing too little new product development. The sound of "Do it! Try it! Fix it!" should echo from the walls of organizations doing first-of-a-kind work. This strategy suggests that the organization utilize the initial products as market probes to collect market information rather than spend excess time and energy trying to predict market response without a product. Simple products that work and communicate concepts to the user win big. Product ideas will dominate; the customer will respond. The do it–try it–fix it cycle moves out of the laboratory and into the marketplace as uncontested profit is generated.

Second is the creation of a "me-too-with-a-twist" product produced at the intersection of the "familiar" column and "established" row; a product type that grows in market acceptance by making steady incremental improvements over its market predecessors with each new release. This is the classification cell where most new product development takes place. The market is known. Organizations try to outdo each other to satisfy the consumer. It's happening with automobiles, refrigerators, televisions, plumbing fixtures, and hand tools. Just about everything that is out there is having something done to it to make it better. In this mode of operation, the flat portion of the curve is significantly shorter in duration. The product's success will be based on its contributions to the market and its time to market.

One of the best documented examples on this type of product is provided by Tracy Kidder in his book *The Soul of a New Machine*. He points to observers in the computer industry making comments such as: "Things change fast in the computer business. A year is a hell of a long time. It's like a year in a dog's life." The case in point follows. Digital Equipment Corporation (DEC) had introduced its VAX 11/ 780, 32-bit, superminicomputer. Data General was caught in the me-too-with-a- twist position. They needed a product of this type to maintain a healthy and robust business profile. The Data General people were saying: "We're gonna get schmeared if we don't react to VAX."

Kidder describes the creation of the Data General Eclipse MV/8000 from a hard- ware and microcode engineering perspective. The product's development began in the fall of 1977. By fall of 1978, its definition had undergone one major revision and staff was being applied. Winter of 1978 saw the product's definition phase reach completion with the partial assembly of two prototypes. No mention was made of a planning phase. The execution phase continued at least through April 29, 1980. On that date, the product had its public introduction. TTM for the MV/8000 was docu- mented to be about 2.5 years.

The MV/8000 has been reported to be a successful product. Many indications were given in Kidder's book that one additional year of development would have made the product not worth doing. Based on these data points, the curve can be constructed to show product success probability as a function of TTM. Product suc- cess remains high for 2.5 years and falls to a low value by the time TTM reaches 3.5 years. It is interesting thinking about what Data General and other firms could do if they designed their organizations to work in this TTM frame of reference as the rule rather than the exception!

As an aid in better understanding this concept in the competitive region, try to synthesize the probability of product success vs. TTM curve for the following ex- amples:

"First-of-a-Kind"	"Me-Too-With-a-Twist"
Apple II	IBM PC
IBM PC	PC clones
Apple MacIntosh	Microsoft Windows
HP LaserJet	LaserJet clones
HP-35 calculator	TI Scientific
Atari 2600	Nintendo

Once a product type becomes part of the "me too" variety, the developer of the initial concept is frequently drawn into me-too-with-a-twist development efforts to retain a firm grip on market share. All of the companies on the above list followed their first-of-a-kind products with me-too-with-a-twist products once competition began to appear.

This transition can be difficult for many corporate cultures to accommodate. Working for incremental improvement rather than creating a breakthrough requires a mind-set that is not resident in many organizations. Turning products over and over sounds dull to some new product developers, but it is fair to say that the bulk of product development occurs by this method. As the years go by, furnaces get better; cars get improved gas mileage; less raw lumber is needed to produce paper; synthetics replace cotton in sports shoes and the prices come down; computers handle more information faster; and so on. With each turn, product development teams attempt to "one-up" their competition. It is a competition between ordinary engineering teams to bring established products to market.

Me-too-with-a-twist product development requires disciplined horizontal and vertical organizational teamwork. Initially, deciding what to do, in a sea of almost endless possibilities, requires the close cooperation between marketing, R&D, and manufacturing. Marketing finds itself drifting in a vast amount of customer, product, and competitive data that is crying out for a simplification algorithm. All too often, R&D can't provide the performance and features that were requested by marketing. Manufacturing is fully aware that present design practices and manufacturing processes will not allow producing the product within preliminary production cost goals. Both manufacturing and R&D say "no way" to the proposed timing. Clearly, a method is needed to focus, in a timely fashion, all of the talents of these team members on the achievement of common group goals as the process moves from definition, through planning, to execution.

Time-to-market performance in this development atmosphere determines future business success. The debate between tandem and concurrent design and process development is over. Product design and manufacturing process design do take place concurrently! As a matter of fact, just about everything associated with the development of iterative products takes place concurrently. New ideas have to get input at the beginning of the new product development cycle to allow for their inclusion into design and processing systems. It's becoming well-accepted knowledge in the cyclic product development environment that a few months' savings on each product turn can produce a six to twelve month advantage by the time the third turn takes place. This time advantage is generally sufficient to gain a dominant competitive position.

The third classification is for "derivative" products. They follow from first-of-a-kind, next-generation, or me-too-with-a-twist products. Any of these three types of product can be modified and used to satisfy the needs of a new consumer. Frequently, derivative products are thought of as A, B, C models that leverage from a standard base. They are different from their hosts in that they are developed for new markets. Strategically, producing products of this type gives the new product development team a distinct advantage—a large percentage of the new product's design and processes already exist.

MIPS is using a derivative product development strategy in their attempt to compete with IBM, Sun, Hewlett-Packard, Motorola, and Intel. It is a small company with about 600 employees and yet the company is behind a product offering that includes integrated circuits, board level subsystems, and full workstations. Just about everything that MIPS does is a derivative of something in existence; a platform that MIPS helped create. At the integrated circuit (IC) level, MIPS designs are fabricated by NEC, Siemens, Sony, and Kubota using state-of-the-art process technology. Board level systems are leveraged from the ICs produced and marketed by the IC manufacturers. Boards that can be obtained from a number of suppliers can be assembled into systems. Even at the system level, MIPS follows a derivative strategy by providing system solutions based on standard compilers, standard graphical user interfaces, standard communication protocols, standard hardware (VME, SCSI, etc.) and the UNIX operating system. The MIPS customer base, which includes Tandem, Digital, Control Data, Pyramid, and Sony, can elect to work with MIPS-supplied solutions at the level of integration that best meets the needs of their market. Thus, users of MIPS products become extensions of the MIPS derivative strategy by providing the final product solution to the end user. Each new offering from MIPS, or each derivative of MIPS designs, attempts to finely segment the market.

Another example is the 747 aircraft from Boeing. Since the A version, it has evolved to the 747–400 model over a period of about 20 years. Each version was designed to meet the needs of a particular market segment as materials and design techniques made it possible.

A characteristic of derivative products is that they are based on a standard platform and changes are made gradually over a period of time to enhance marketability. Time to market for derivative products is usually a fraction of that that was needed to create the platform.

In making derivatives of platform products for test and measurement market segments, I have observed TTM numbers in the range of six months to three years. The shorter TTM figures correspond to situations where competitive pressures were the highest. In particular, we were in competition with other vendors to supply units under a competitive bidding and delivery process set up by the Department of Defense. Each product upgrade was surrounded by its unique set of circumstances that determined its success in its intended market. TTM wasn't the only factor, but it sure was high on the priority list! Obtaining information from the market to synthesize the success-vs.-TTM curve for each situation would have been a straightforward process.

Lastly, consideration is given to next-generation products. These are meant to replace an aging product family that has little room remaining for market growth opportunity and that is being picked apart by the competition. For these products, TTM is of prime consideration. It should not be so short that new products appear before they are needed nor too long that products reach the market too late.

Care must be taken to offer new capabilities needed and not just supply a smaller, faster, and less expensive version of the original. The next-generation products need to be in tune with the overall market, not merely redesigned versions of their predecessors. In all cases that I've studied, TTM was longer and engineering effort (number of person-months) was higher on the next-generation product than it was on the first-of-a-kind effort. This suggests that a good TTM starting point estimate on a next-generation product should at least be equal to the TTM of the product about to be replaced.

Often, improved competitiveness results from next-generation efforts moving the entire organization in a new direction. Xerox used a next-generation program to reduce dependence on a large number of material suppliers and establish new manufacturing capability. Hewlett-Packard is using global technology throughout all functional areas as a strategy to multiply its business effectiveness in its laser printers. IBM, Texas Instruments, NCR, and Apple have used next-generation efforts to integrate, fuse, and develop design for manufacturing (DFM) and computer-integrated manufacturing (CIM) processes and systems.

Generally speaking, describing next-generation products is a relatively easy task, since prior product descriptions are already in existence and data has been collected from the marketplace over a period of years. Similarly, knowledge about customer desires and the competition is the highest for this type of product. Next-generation products are frequently "pulled" into the market. Knowledge from the market is supplied to the product development process to create a product that meets the requirements set by the market. It is usually very clear to the development team what has to be done, but usually not too clear how to achieve project objectives. For this reason, next-generation efforts tend to produce major gains in the tactical and logistical performance of an organization.

If you use this TTM concept, you will be able to access data that allows for the prediction of the flat region of the curve in your particular business and competitive environment. In general, internal data is relatively easy to collect and analyze based on product financial records. These records usually contain investment and sales figures. Investment data can be used to extract the TTM of a given product. Sales information can be used to measure the product's market success. Analysis of similar products within similar markets should produce information that is very useful to correlate product success with TTM. Also, making estimates on competitors' performance, while not an exact science, can be done with sufficient accuracy to determine a few of the early breakpoints. This exercise puts a numerical value on TTM that an organization needs to meet or beat to remain competitive.

The last, and possibly most important, reason for giving attention to TTM is the financial leverage that can be obtained. *Fortune Magazine* reported in their February 13, 1989, issue that, in an economic model developed by the consulting firm of McKinsey and Company, high-tech products that come to market six months late but on budget will earn 33 percent less profit over five years. In contrast, coming out on time and 50 percent over budget cuts profits only 4 percent. *Think of the leverage that can be obtained by coming out on time and within budget!*

Figure 1.4 introduces concepts that help link delay in development time to sales revenues that are lost as a result of the delay. The figure and the supporting discussion are intended to help explain, from a financial perspective, the necessity of managing time to market within the new product development process.

Two negative cash flow profiles are supplied for the investment phase of the product development effort.

The "on time" monthly expenditure of cash closely follows the recommended staffing profile discussed under the control of development expenses. This profile makes it evident that the development effort went through a complete definition phase that was followed by a short, but detailed, planning phase before the new

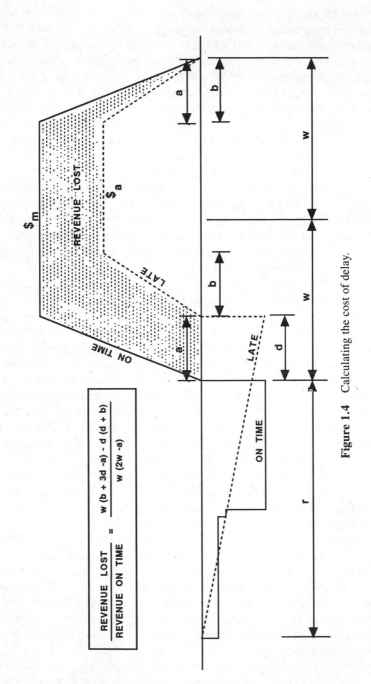

$$\frac{\text{REVENUE LOST}}{\text{REVENUE ON TIME}} = \frac{w(b + 3d - a) - d(d + b)}{w(2w - a)}$$

Figure 1.4 Calculating the cost of delay.

product development effort was executed. Total development costs for the effort are obtained by finding the area under the curve. As was previously emphasized, this staffing profile has the desirable characteristic that less than 20 percent of total expenses occur prior to full staffing and resulting commitment to the completion of the new product development effort. Also, note that the definition and planning effort consumes about 50 percent of the total on-time development time r. The remaining 50 percent of development time and 80 percent of cash are consumed during the execution phase. As can be seen from the figure, this development effort yields a product that enters the market window w at time zero, the point where delay d is equal to zero.

The "late" profile is represented by a triangle. Total expenses, as computed by finding the area under the curve, for this profile are drawn to be approximately equal to the on-time profile. Setting the areas under the curves to equal each other is a scenario that is generous toward the "ramp" profile. It is not unusual to have new product development efforts that take longer cost more. This triangular expense profile is quite typical of projects that tend to have delayed market entry. In this case, on-time development time r is extended by d. Total development time then becomes $r + d$ in duration as the product enters the market well into its market window w.

Many mechanisms are at work to produce the triangular shape. During the definition phase of the product development effort, a few individuals may work in isolation from others and without participation from key individuals and customers. As the product's definition is communicated to more people and across functional boundaries, it tends to undergo a substantial amount of revision. Where the definition began and where it ended becomes ambiguous. All that is known for certain is that the idea for the product got a start and that staff and expenses grew over time. Somewhere in this amorphous period, a schedule usually appears. It will be based on the most optimistic factors known about the new product development effort. No attempt is made to take into account historical information to help predict the future. The development team moves forward based on a view that is far removed from reality. As problems are encountered, adjustments are made to the product's definition and its very ambiguous plan. These adjustments almost always require further resources, expenditure, and time. Actions of this type tend to increase the monthly rate of cash expenditure and extend development time; in effect, the "late" triangle continues to extend in the x and y directions. Eventually, enough work and rework will have gone into the new product development effort to make the product ready for customer delivery.

For discussion purposes, the assumption is made that either of these profiles is possible for the new product under development. This assumption then makes it possible to explore the product's revenue-generation potential as a function of its market entry delay.

Estimation of the product's maximum sales life, $2w$, becomes the starting point of the exercise. This life is defined as the time from the product's expected availability for sale to the time when sales diminish to approximately zero. It is a number that is best arrived at using a consensual decision-making process within a group that understands the product and the market.

2w figures tend to scale with product classification. Me-too-with-a-twist products range from a one- to a four-year period. Next-generation products extend from two to six years. A successful first-of-a-kind effort can have a minimum sales life of three years and a maximum of eight, or more, years. Derivative products are distributed over this entire range depending on product and market considerations. In general, the more that is known about the market, the greater the certainty in the determination of 2w. It is not necessary to spend excess time in making this estimate. A group effort that produces a result that feels accurate to within 30 percent is completely satisfactory.

The market window w is defined as the maximum sales life divided by two. A product that begins its sales life at the start of its market window is expected to sell more over its sales life than one that enters the window at some later period. Likewise, a product that enters the market near the end of the market window can be expected to produce little or no sales return. The peak monthly sales realized when a product enters the market on time, m, is scaled by the relationship of $(1 - d/w)$ to arrive at the peak revenue generated, a, when the product enters the market with d being greater than zero. The model now has the ability to scale revenue as a function of delay into the market window.

Growth in monthly sales is set with the factor a for on-time market entry and with the factor b for delayed market entry. These factors allow for the establishment of the growth, maturity, and decline aspects of the product's life cycle. a and b are assumed to be equal for the growth and decline segments of the life cycle.

On-time total revenue is found by determining the area under the on-time curve; similarly, total late-market-entry revenue is found by calculating the area under the late curve. The total revenue lost by being late is found by subtracting total late revenue from total on-time revenue. Once this is done, it becomes possible to compute the ratio of revenue lost to revenue on time:

$$\frac{\text{Revenue lost}}{\text{revenue on time}} = \frac{w(b + 3d - a) - d(d + b)}{w(2w - a)}$$

where

w = market window
d = delay in market entry
a = time to rise to and fall from mature on-time sale
b = time to rise to and fall from mature late-market-entry sales

The model was used to explore R_l/R_{ot} sensitivity to various sales profiles and for each profile determine ratio sensitivity to delay d. To simplify analysis, market window w was normalized to one and delay is expressed as a percent of market window. Those inclined to do the detailed analysis will produce the following results:

Delay, %	R_l/R_{ot}	Error, %
0.0	0.00	0.0
10.0	0.15	10
20.0	0.30	9
30.0	0.43	8
40.0	0.54	7

The error column gives an indication of the variability of the R_l/R_{ot} figure for various sales profiles and shows that the ratio is minimally dependent on the shape of the sales profile; that is, once agreement is reached on the likely shape of the on-time profile and if this profile is maintained as d is varied, the R_l/R_{ot} resultant is almost entirely a function of delay. This is a significant characteristic of this model that can be put to good use in determining the product's sensitivity to delay.

While the introduction to the model may be conceptually and mathematically detailed, utilization now becomes a simple exercise. To become sensitized to the cost of delay for a particular new product development effort, determine the market window w or its maximum sales life $2w$; estimate total revenues from the project if it were to enter the market at a delay of zero; compute delay as a percentage of market window; and calculate revenue lost by using the information in the above table for various delay percentages.

As an example, consider a project with the following estimated parameters:

On-time sales life	4 years
Market window	2 years
Total on-time sales	$100 million

Simple calculations produce the results below:

Cost of being 10% late (2.4 months)	$15 million
Cost of being 20% late (4.8 months)	$30 million
Cost of being 30% late (7.2 months)	$43 million
Cost of being 40% late (9.6 months)	$53 million

A ballpark number for the cost of delay on this example product development effort is $5 million per month for each month late. When being on time has this much financial benefit, who would reject spending a few more development dollars, adding additional resources, and adjusting product attributes to get the work completed?

This model was developed to illustrate cost of delay concepts and to be relatively easy to use by various members of the new product development team. Its application should facilitate the identification of critical time-dependent variables for a particular product development effort. Should a more detailed model be desirable, the detailed discussion in the Appendix can be referred to (see "Time-to-Market Cost Model" therein). Also, the interested reader may want to refer to the works of Don Reinertsen to develop models that offer a higher degree of sophistication.

Being convinced that it is important to be on time and being on time are two entirely different issues. The former can be simply a mental exercise, whereas the latter requires hard work, discipline, and the development and application of an effective new product development process. The material offered in this text is intended to document an efficient and timely methodology for the realization of new products. All the process parameters described in future chapters are optimized for speed of execution while preserving complete process quality.

1.2.6 "Can Do"

An effective new product development process must do all it can to preserve and nourish the "can do" spirit of the participants. This human quality is an absolute requirement for process success. Optimism is a must to be a winner in business. Experience shows that this attitude will grow and develop over time if people

Think highly of themselves.
Perform their work because they are internally motivated.
Strive to realize a driving vision.
Establish short-, intermediate-, and long-range goals.
Develop a realistic appraisal of progress toward achievement of their goals.
Take time to have fun.

The process contained in this book attempts to address the above demands. Self-respect is encouraged by soliciting input from process participants over time to give them control over the process. Internal motivation is maintained by minimization of process "bureaucratic" structure and facilitation of rapid decision making by those closest to the problems. The time flow of work over the definition, plan, and execution life cycle was created to build a team vision, develop plans, set goals, measure performance, and provide timely feedback to process participants. On the surface, this prescription for success sounds too simple. As we go forward, you will learn that the challenge is to keep these simple concepts from getting lost in excess process information detail.

As for the fun element, the process will allow you to think that the world is about to end, but will remind you that it just might look better in the morning. It encourages you to get mad and then does a few things to help you get over it. It makes it easy to separate your ego from your position. You will be allowed to make your own choices (for a few, you'll wish you could blame them on someone else). Controls are in place to remind you to check up on little things. You will be constantly reminded to stay calm. Above all, this new product development process will teach you (if you don't already know) that *optimism is a force multiplier*!

1.3 PROCESS PROBLEMS

As attempts are made to make the process move faster, as pressure is applied to reduce spending, as change creates more complexity and elevates conformity issues, while it becomes progressively more difficult to visualize the new product development process and the product it is supposed to produce, it is likely that mistakes will be made and resulting problems will be detected. Surprisingly, the problems that significantly impact product development form a finite list:

Shortage of engineers	Organizational instability
Skills not matching job	Inexperienced staff
Inadequate progress reviews	Dependent project delay
"Creeping" features	Shift in market requirements
Unanticipated competition	Technology changes
Testing reveals problems	User requests modification
Customer requests change	Shortage of design tools
Shortage of resources	Design errors
Unforeseen tasks	Internal miscommunication
Additional time needed	Retest reveals problems
Customer miscommunication	Long turnaround time
Testing turnaround time	Change in marketing plan
Acts of God	Management methodologies
Leadership methodologies	Program is too big
Inadequate planning	Doing it the first time
Poor product definition	People turnover
Overworked/burnout	Internal competition
Lack of management support	Money runs out
Changes for manufacturing	Product looking for market
Sales forecast change	Difficulty communicating concepts
Excessive planning	Ineffective decision making

The preceding list was assembled from firsthand experience in leading new product development efforts and by collecting feedback from an assortment of project/program managers that were either work associates, clients, or students.

Project and program managers have superior insight into new product development process problems. They have been asked to be the "champions" of the process by their organization's management team. Most, if not all, of these individuals have had experiences that were excruciating. Much of their work was delivered late, exceeded expected production costs, and failed to meet full performance requirements. Their recollection of what happened and why is of much substance. They tend to view the process across functional partitions and frequently must work with all levels of the functional management structure. These individuals usually have an excellent horizontal and vertical process view within their operating environment. Their collective voice of experience is accurate. For this reason, attention is focused on the factors these managers cited as being the source of serious problems.

Examination of the list reveals no items that cause the new product development process to move faster, spend less money, or make a better product. On the contrary, it points to some rather severe new product development process imperfections. Analysis of the list led to the realization that the problems could be grouped by process phase; that is, groupings could be established for the definition, planning, and execution phases of the new product development effort.

The problems that can be segmented in the definition-related category are

Poor product definition
"Creeping" features
Ineffective decision making
Difficulty communicating concepts
Customer miscommunication
Internal miscommunication
Lack of management support
Shortage of design tools
Shortage of engineers
Shortage of resources
Changes for manufacturing
Money runs out
Product looking for market
Program is too big
Technology changes

Subclassifications of this grouping include product definition, communication, resources, and others.

I have vivid memories of developing several new products whose definition was either incomplete or unstable. The problems associated with poor performance in this area remained with the product development effort to the end. From a product definition perspective, accurate generation of information about what is to be done is a major issue for the new product development team. The product must be defined! A definition must be done in such a way that it can be evaluated to produce a measure of its quality. After evaluation, the definition must retain a degree of stability. Decisions that lead to the production of the product's definition must be timely and accurate. If features are allowed to "creep," the work to produce the new product expands, thus resulting in the continued consumption of available resources and time expansion of the project.

Of course, the definition must be communicated to the entire product development team. There are many opportunities for communication error. Communication links must extend from the customer to the product developers and between product developers. It's all too common an occurrence for a key individual on the human network to miss the reception of critical information or to interpret existing data in an incorrect fashion. Links between the management team and the development team require error-free transmission. Communications extend beyond the transfer of requirements information. Frequently, concepts and "gray"-area data needs to be disseminated and refined. Miscommunication of concept and product information has serious product development consequences.

As preliminary information about what is to be done surfaces, thought processes shift to thinking through how they might be accomplished. Often, in this early phase of product conceptualization, errors are made in estimating tools, resources, and engineers needed to develop a product. Estimation errors then lead to projects that expand to fill the void between optimistic thinking and reality. With the passage of time, the product development effort can become too large or just grow to such an extent that available funding is consumed and the venture is terminated.

Even if the definition phase of a product development effort was executed in a fashion such as to avoid making errors, the planning phase may not have been. Planning errors are made evident when the following type of problems are detected:

Inadequate planning
Excessive planning
Inadequate progress reviews
Doing a job for the first time
Skills not matching job
Additional time needed
Long turnaround time
Testing turnaround time
Internal competition
Management methodologies
Overworked/burnout
Management-driven schedules

Doing too much or too little planning correlates with knowing when to produce a plan and how to produce a plan. When to initiate the planning effort depends on the quality and stability of the product's definition. Obviously, attempting to create a plan based on an unstable definition is an exercise in futility. The plan will be forced to track a rapidly moving target requiring frequent and extensive plan modification. However, it's unreasonable to expect that a definition will forever remain constant. Therefore, identification of the point in time when the rate of change of the definition is small enough to allow the production of a quality plan becomes a critical new product development process issue. Once the optimum time has been identified, then technique becomes important. Plans must be swiftly produced on a shared and distributed basis to develop the necessary insights about the reality of developing the new product while using a collective process to enhance communication and teamwork. Shared planning techniques build commitment at the foundation of the plan.

Teams that know when to plan and how to plan can then develop the additional skills needed to determine the right number of progress reviews as a milestone sequence is produced for the project. What might be thought by some to be an elementary concept can have profound implications on the overall effort. The number of progress reviews needed by an experienced product development team creating me-too-with-a-twist products over a relatively short development time frame could be significantly different from the number and type needed for a first-of-a-kind effort that takes place over a multiyear time scale. Development of the appropriate timeline and milestone sequence for the project is only the beginning.

Attention must be given to developing a plan with just the right amount of detail. Is it necessary to define activity structure down to a daily level of specificity for each person on the team or will a framework that looks at a task, or two, a month per individual suffice? Likewise, it is likely that the structure of the plan will be influenced by the experience level of the team assembled to produce the new product. An experienced team may know full well the detail behind cross-functional interactions, whereas a new team may require crisp direction and documentation about what they are to give, to whom, and when. For a group of individuals doing something for the first time, a quality plan may be the only vehicle that eliminates, or minimizes, sins of omission while forcefully guiding the effort away from sins of commission. Creation of the distributed plan must be sensitive to the experience level and size of each work group to ensure that the right balance is achieved.

Lastly, the planning team needs to become skilled in making accurate estimates of work and effective at making resource assignments. Without these skills, the work done on the plan to get it to this point is of limited value. A plan is not complete unless individuals know when they are expected to start and complete their assignments—this naturally implies that individuals know their assignments.

As a final safeguard from entering the execution phase prematurely, the planning phase can be used to measure the stability of a product's definition by attempting to produce the plan. If a quality plan, as verified by thorough analysis, cannot be produced and retained for a brief period of time, a serious new product development process problem has been identified. Once alerted to the existence of the problem, actions can be initiated to determine cause and develop corrections.

By this point in the evolution of a typical new product, it is almost certain that serious problems exist within the development process, the product, or both, but the effort is certain to move forward and enter full-scale development. Once this occurs, no new product development undertaking is exempt from discovering and correcting problems—detection and elimination of problems is a principal reason for the execution phase in the first place. Those problems that are to be detected in the execution phase of the effort are likely to be produced by

People turnover
Organizational instability
Inexperienced staff

Shift in market requirements

Change in marketing plan

Sales forecast change

Design errors

Testing exposes problems

Retest reveals problems

Customer requests change

User requests modification

Dependent project delay

Unforeseen tasks

Unanticipated competition

Acts of God

Technology changes

Leadership methodologies

People will be added to and subtracted from the effort over a period of time. The longer it takes to complete the execution phase, the more susceptible the new product effort becomes to turnover and organizational instability. Some efforts take too long because they take too long! It's relatively easy to determine the number of personnel changes that will have to be accommodated over a year just by computing the average turnover rate of the department/organization. A project staffed with 50 people that is operating in an environment that has a 20-percent annual turnover rate is likely to see 10 people depart over the year. While it's easy to predict quantity, it's almost impossible to predict the individuals. Resources and processes must be in place to rapidly adjust to changes of this type.

The next group of three items, shift in market requirements, change in marketing plan, and sales forecast change, relates to less than perfect performance from marketing individuals. It's reasonable to expect that some change in these areas will occur over the execution phase. As previously mentioned, the longer the execution phase extends, the higher the probability that problems, produced by causes of this type, are likely to be produced. Expect some information produced by marketing individuals to change and some to be in error. The team working on the product development effort needs to know how to rapidly make the adjustments as error is detected and change occurs.

Just like their marketing counterparts, R&D personnel make mistakes and change their minds. In their case, the problems they create are usually exposed by testing. During test, serious performance deficiencies and troublesome product "bugs" are discovered. Time and resources must be allocated to the determination of cause-and-effect relationships and the eventual solution of problems by the design team.

As the execution phase nears completion, the product is oftentimes supplied to internal users and selected external customers for evaluation and feedback. Users and customers tend to be very generous with the critical comments they have on the product. Frequently, their input will be so strong that portions of the product will have to be reengineered. The new product development team will have to evaluate feedback objectively and take appropriate action. As is the case with most execution phase problems, obstacles of this type can be anticipated. The team knows of their occurrence—it does not know when and how the strike will take place, but, it can set aside, in advance, the time and resources to respond.

Unforeseen tasks and dependent project delay usually arise because work that could have been planned wasn't and because lateness is often due to a dependency on another, often unanticipated factor. Dealing with issues of this type is a common occurrence for those working in manufacturing-related disciplines—it is a matter of identifying the flow of deliverables and scheduling their timely delivery. Just-in-time (JIT) theory and techniques apply. The tactic is to keep running the new product development process over and over to generate the opportunity to make incremental improvements each time it is utilized. It's difficult, if not impossible, to create a perfect "one shot" process; process history must be recorded, measurements made, and data analyzed so that they can be applied to future new product development efforts.

The more exposure you have to the development of new products, the more familiar you become with the items on the preceding lists! It becomes a matter of utmost priority to create a new product development process that navigates around these traps and ultimately gets a quality product to market in a timely fashion.

1.4 ORGANIZATIONAL MODELS

Modeling the organizational structure in which a new product development team operates is the next topic of discussion. Teams perform their work within some type of structure that provides a hint of what participants are to accomplish. If somewhat successful, the structure will partially define a number of important relationships for each member of the new product development team. The initial core of the structure is built around the concept of a business—the purchase, transformation, and sale of goods in an attempt to make a profit.

Surrounding the core of the business are people, processes, and products. People make the decisions on what to buy, what to produce, how to do the production, where to do the production, how to distribute output, and how to collect revenue. In a small business, decisions of this type are made by a few individuals and work can progress. Success will bring about growth in the size of the organization. As this growth takes place, the simple nucleic business core will become hierarchically decomposed. This decomposition usually results in the creation of a functional organization that now consists of individuals working in research and development, marketing, and manufacturing. Continued success continues the segmentation of functions into subfunctions. To keep functions and subfunctions working as a team, a management structure is added to the organization structure to keep the entire system functioning. Management will then deal with the strategy, tactics, and logistics needed to keep the business in existence.

Business managers, usually with help from key advisors, must then make enough sense of their business environment to keep the business in a profit-making position. These people will have to understand why they are in business and where they want to take the business over time. Knowledge of the products and services the business must provide is vital. Ideally, product and service offerings should be tightly coupled to customer needs and desires, implying that business managers should be tightly coupled with their customers. Information on the marketplace and the competition must be collected, analyzed, and acted on on a periodic basis. As ideas for new products surface, skills will be needed to select the ideas that offer the highest probability of continued business prosperity. Decisions regarding which new product ideas to pursue and which ideas to leave undeveloped will become critical determinants of future success. Knowing how to assess present business conditions accurately, running the business as it happens on a real-time basis, and developing new products for future success are the primary challenges facing most any viable business operation.

Figure 1.5 shows a business modeled from a holistic and integrated perspective. As the figure is scanned radially outward, the overall business segments into people, processes, and products. These categories are then further functionally decomposed into manufacturing, marketing, and research and development. Each member of the research and development, marketing, and manufacturing "triad" then maps their perceptions of the environment into a comprehensive strategic, tactical, and logistical business framework. Coupling the business organization to the environment is accomplished by rotating the business pointer to a particular environmental factor. In the example, the environmental factor selected is "time to market." With the pointer in this position, one can begin to evaluate the effects time to market has on the organization. Strategically how do marketing, research and development, and manufacturing respond with their products, processes, and people? What tactics do each of the functions employ? And finally, how will strategy and tactics affect the logistics of each function?

As you can see, rotating the pointer to each environmental factor on the circumference will produce a wide range of responses from the organization. An organization that understands itself and its environment extremely well can rotate the pointer to any environmental factor and accurately describe what each of its functions is doing along strategic, tactical, and logistic lines to improve its competitive business position through products, processes, and people. When an overlooked area surfaces, those utilizing the model can form a conceptual bridge to the solution of the problem. A model of this type aids in systematic review and utilization of information in an understandable form.

Another organization having difficulty with the "radar" model might choose a more conventional analytic approach. Obviously, one way to capture the response at each pointer position is to transform the "radar" organization model into a matrix representation. To perform the transformation, designate the 32 environmental factors as column headings and the 27 possible combinations of categories within the business triangles as rows, which results in the creation of an 864-cell model. As you might expect, the data contained within each cell has the potential to be rather

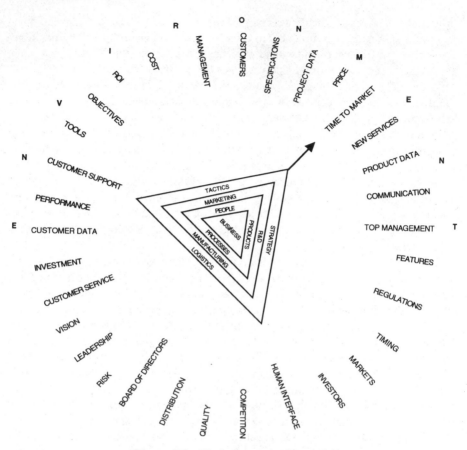

Figure 1.5 "Radar" organization model.

voluminous, resulting in continued geometric growth of the model. It doesn't take much more expansion to realize that a single-level matrix implementation of the "radar" diagram becomes overly complex as it loses its utility and visual impact.

To overcome the problems associated with organizational modeling within one level, hierarchical decompositions and matrix linkages are often employed. First, it is important to recognize that many relationships can be modeled in two dimensions and that usually an axis from one two-dimensional model can be made to link with an axis on another model (like in the game of dominoes). This is one of the simple beauties of matrix models!

For instance, consider a common market model representation that has product types plotted along the *x* axis and customer groups plotted along the *y* axis. This model is given in the upper left-hand corner of Figure 1.6. For discussion purposes, the model takes a simple view of the printer market and maps product descriptions into the market. As a stand-alone structure, it can be very useful. With the use of numbers or symbols at each intersection, an indication of how well a product is

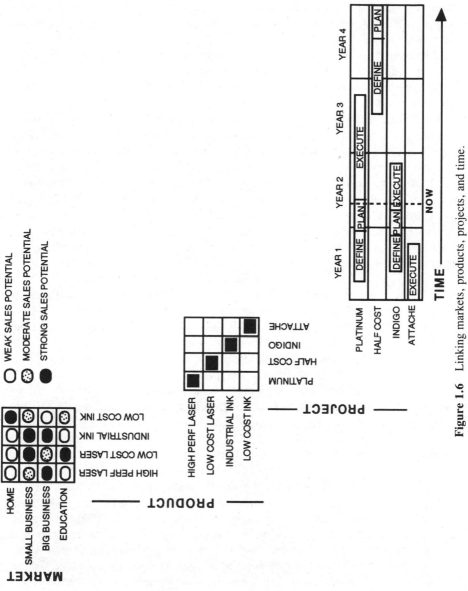

Figure 1.6 Linking markets, products, projects, and time.

expected to perform in the market can be provided. The figure shows this interaction between product and market symbolically. The market for printers is divided into four segments: home, small business, big business, and education. Four product types are arranged orthogonally to the market segments. At each intersection, a symbol is recorded that is used to indicate how well a product is expected to perform in a given market. For example, a low-cost ink printer is expected to do extremely well in the home market, have moderate sales potential in the small business and education markets, and have very weak potential in the big business arena. The same thought processes are followed on the remaining 12 cells. Once complete, product information is then used to form the y axis of the next matrix to link products to projects.

Ideally, the second matrix will only have entries along the diagonal. Entries appearing off the diagonal indicate that multiple project teams are working on the same classification of product. Most of the time, product development overlap is a situation that an organization will want to eliminate once it becomes visible. On the other hand, there may be times when it is desired. This matrix view brings the situation to the full attention of decision makers. A missing entry on the diagonal indicates that a product alternative has not been elevated to project status.

Lastly, form a matrix that has project on the y axis and time on the x axis. This view now gives individuals within the organization the information they need to get information about new product development timing between multiple projects within the organization. The matrix in the lower right corner of Figure 1.6 furnishes this view. At a glance, members of this organization know the number of new product development efforts currently under way. At any given instant, it is a simple matter to identify where any project is in its life cycle by identifying where the "now" line intersects each project's time line. Additionally, a view of the past and the future is provided to simplify comprehension and aid in the formation of associations and relative characteristics between projects over a period of time.

As can be seen, some simple concepts linked together created a coupling between the customer, the products being produced, the people producing the product (projects), and the life cycle (time) to get the product to market. The importance of establishing these simple and visual relationships cannot be overestimated. Each matrix can be viewed independently; and yet, for those in need of connectivity and transformation information, it is readily available for consumption. A modeling technique of this type is midway between the conceptual "radar" model and the completely analytic 864-cell single-tier matrix model discussed earlier.

Quality functional deployment (QFD) utilizes the technique of matrix linking to provide an organized way of ensuring that the customer's voice is properly captured and used to define, design, and manufacture a product. In its most elementary form, a matrix that links customer requirements to design requirements is produced by the means that are most suitable to link a particular design team with a representative sample of customers. Once this matrix is complete, the design requirements axis is linked to another axis labeled "parts design." Design requirements are then used to appropriately design each of the parts that goes into the end product. As information on this matrix nears completion, the parts design axis is coupled to the manufacturing process axis. Part design and manufacturing process are precisely optimized for the product. Completion of this transformation then provides the link for the manufacturing process axis to be linked to a build quantity axis that was produced by another group doing sales forecasts. Figure 1.7 illustrates these concepts and demonstrates the strengths of this technique. The potency of the visualization techniques becomes obvious. Also, the images and data begin to stimulate thoughts on other matrix links that could be formed to couple other functions' work into the linked matrix model. Note, however, that while the technique is more visually appealing than a large nondecomposed matrix is, the total amount of information being managed through matrix linking is the same. This fact places a limit on how far an organization may want to explode detailed information in this fashion.

To be effective aids, the matrices must model reality in a way that is reasonable and makes good business sense. At the same time, the model must avoid excess

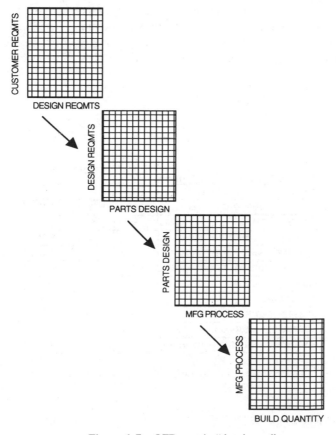

Figure 1.7   QFD matrix "dominoes."

complexity to retain its complete usefulness as a tool and decision-making aid. More elements about the effective utilization of matrices as modeling tools will be discussed in future chapters.

From the perspective of those working on the new product development effort, they will participate on teams that will be working on definition, planning, and execution as a function of time. The output they produce is linked to the market through the product that is delivered for sale. To better understand how each function's contributions extend to the end product, it is instructive to construct a more detailed matrix using the techniques described in previous paragraphs. The new matrix is shown in Figure 1.8.

This matrix is simply an extension of the last matrix within Figure 1.6. In this case, the time axis, represented by define, plan, and execute for each project, becomes the x axis of the new matrix. The y axis is formed by the functional decomposition of the organization that will contribute to the realization of the new product.

After studying the functional composition of such companies as Hewlett-Packard, Intel, National Semiconductor, Ford, Boeing, Goodyear, Sony, U.S. Robotics, Nike, Actel, and a number of other firms, a common functional structure began to emerge. Those functions that were found to exist in all the companies studied are listed along the vertical column in Figure 1.8. Nomenclature differs

	DEFINE	PLAN	EXECUTE
PROGRAM/PROJECT MGMT TEAM	S	S	S-M
DOCUMENTATION TEAM	S	S	M-L
PRODUCT SUPPORT TEAM	S	S	S-M
SALES TEAM	S	S	M-L
PRODUCT MARKETING TEAM	S	M	S-M
QUALITY ASSURANCE TEAM	S	S	S-M
MANUFACTURING TEAM	S	M	M-L
FINANCE TEAM	S	S	S
PERSONNEL	S	S	S-M
SYSTEM ENGINEERING TEAM	M	M	M-L
SOFTWARE ENGINEERING TEAM	S	M	M-L
HARDWARE ENGINEERING TEAM	S	M	M-L

[APPROXIMATE SUMMATION ⊥ <10 ⊥ >15 ⊥ >25]

S = 0 TO 2 PEOPLE   **M** = 3 TO 5 PEOPLE   **L** = > 5 PEOPLE

Figure 1.8   Work function staffing vs. time.

slightly from company to company, but the work performed by the work teams is essentially the same. At first, the consistency between companies was somewhat surprising, but after added review of the information collected, I became comfortable with the similarity and decided to put this discovery to work in a general fashion within the models I develop.

Not all product development undertakings require all teams. Some teams absorb the work of others. Most often, teams are called upon to do their regular jobs while supporting the introduction of new products. It is usually up to the project manager to capture the attention of these teams and get them to perform the tasks necessary to complete the subproject for which they are responsible.

It is clear to see, as a project moves from definition to planning to execution, that the number of people involved grows. The functions that start with small staffing and tend to stay staffed lightly over the duration of the new product development effort are project/program management, quality assurance, finance, and personnel. Depending on the complexity of the product under development, all other functions have the potential to increase staffing to relatively large numbers. It is not at all unusual to see an effort that starts with a small staff grow to become a major consumer of human resources. In observing new product development efforts across a wide spectrum, not one was found that was successfully executed without the contribution from at least 20 or more people. This observation suggests that there is just no such thing anymore as a small project! The financial implications of this observation were previously covered in Section 1.2.4. Future chapters will lend insights on blending staff growth into the modeling, measurement, and analysis of the new product development process.

The next organizational view that will be discussed is that of the hierarchy. Its structure and graphical representation (Fig. 1.9) are probably familiar to most new product development process participants. Frequently, project managers get a cold chill just by counting the number of dotted-line relationships that are needed to execute their job responsibilities. Figure 1.9 is a point of reference. It is not suggested that one spend much time and energy attempting to change it. It has been proven by a countless number of new product development process participants that it is possible to reach a state of acceptance and get on with the job knowing full well that most hierarchy charts don't tell what is really happening.

Finally, Figure 1.10 presents a structure described by Harvey Levine in his book *Project Management Using Microcomputers*. This structure is called the matrix organization. In it, the hierarchical organization is somewhat preserved, but a new column (new products) is added to clarify the structure and the roles performed by individuals within the organization.

In this structure, an organization creates a function that is responsible for the management of all projects/programs in addition to the regular functions of R&D, marketing, manufacturing, and support. (I'll leave it as an exercise to describe the matrix transformation that was used to produce the new model.) The roles of the functional managers take on relatively clear definitions. Functional managers, with the exception of the new products manager, work to optimize their teams to be

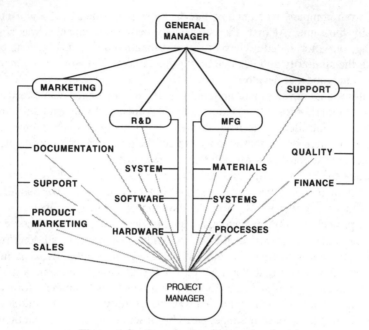

Figure 1.9 Organizational hierarchy chart.

GENERAL MANAGER																
NEW PRODUCTS MANAGER	R&D MANAGER				MKTG MANAGER				MFG MANAGER				SUPPORT MANAGER			
	SYSTEMS	SOFTWARE	HARDWARE	OTHER R&D	DOCUMENTATION	PRODUCT MKTG	SALES	OTHER MKTG	MATERIALS	MFG SYSTEMS	PROCESSES	OTHER MFG	QUALITY	PERSONNEL	FINANCE	OTHER
PROJECT K			X		X	X	X		X	X	X		X		X	
PROJECT T		X		X	X	X	X		X		X	X	X	X	X	
PROGRAM C	X	X	X		X	X	X	X	X	X	X	X	X	X	X	X

Figure 1.10 A project/program matrix organization.

highly effective along the column to service many projects. The new products function's focus is along the horizontal. Each project is headed by a project manager who is responsible for making the right things happen within each cell of the matrix that relates to their assigned project.

In its elementary form, the structure is designed to function with a minimal amount of interaction between individuals to arrive at time to market (TTM), project priorities, resource allocations, and budgets. Each project/program (K, T, C in this example) manager works with managers and team members along the row to define, plan, and execute the project/program. The orchestration of all projects operating concurrently is the responsibility of the new products functional manager. This manager will work across the functional management row to ensure that the needs from each project have been accurately communicated to and effectively addressed by each of the functional managers. The general manager is the overall process observer who ensures effective operation of the project matrix.

In practice, the role specified to be performed by the new products functional manager is frequently delegated to the project managers, thereby eliminating the new products functional management position. In this form, the individual project managers now deal directly with the team, team managers, and functional managers. By default, the job of the new products functional manager is absorbed by the organization in an unstructured and undefined fashion. Organizations that cannot afford to operate the new products functional management position should give consideration to assignment of this job responsibility to one of the existing functional managers. This strategy greatly reduces the work load on the project manager, allowing he or she to interact with a number of individuals horizontally and only a few individuals vertically. If project managers are primarily based in one function within an organization, it would be reasonable to have that functional manager serve as the new products coordinator.

At this point, it is important to examine the organizational models that are used within your organization and compare them with the material that has just been presented. This examination accomplishes four important objectives:

1. You can begin to gain an understanding about how to go about modeling your organization's new product development process.
2. You will cultivate an insight as to who the people are that manage the process.
3. You will start to develop a more focused view on the output items that other groups contribute to the overall effort.
4. You can begin to identify areas for organizational improvement.

Performing the exercise would seem quite simple on the surface. For those who follow the suggestion, you are sure to increase your understanding of your job function and expand your awareness ("maze brightness") of your working environment. This added insight will help in relating to, adopting, and applying the information contained in subsequent sections of this book.

1.5 PATHS TO SUCCESS

It is possible that work under way now will soon yield a magic formula that will make new product development a routine process. Efforts are being made around the world to build expert systems that theoretically arm average people with substantially above-average capabilities. Within the United States, the National Science Foundation (NSF) is sponsoring research on a topic referred to as the computer-integrated enterprise (CIE). NSF believes that improvements in engineering and technology management could significantly enhance industrial productivity. The possibility exists that a system could be produced to fully model the new product development process and produce tools and procedures for having computers and artificial intelligence systems manage the creation of new products. As a system of this type appears, it is apt to contain gigabytes of artificial intelligence. Quite likely, the software to run the system would have been produced using new high-level languages, object-oriented programs, or graphical programming techniques. The computational machines themselves will probably have their electronics produced on silicon or gallium–arsenide compilers. We'll probably call the machines that embrace us with this collection of hardware and software "super workstations." Without doubt, these high-intellect and high-capability tools will have to be networked to allow them to achieve their maximum potential. Such tools will hopefully appear early in the next century. For the present, a more down-to-earth approach will be discussed.

There are a few actions that can be taken now to assure new product development success. A list of ten follows:

Winning Ways of Doing New Product Development

1. Make new product development a controlled process.
2. Keep an eye on the world.
3. Involve all relevant people from the start.
4. Assemble and act out information in concert.
5. Have a representative object of the end product in view.
6. Learn how to make decisions quickly.
7. Work with competitive tools and methods.
8. Entrust execution to competent people.
9. In the event of problems, adjust only the affected areas.
10. Maintain the "can do" vitality in the organization.

Without doubt, new product development can be a systematic and controlled process. The process can be defined. Information exchanges between people in different functions can be determined and communicated. Output as a function of time can be specified. The quantity and quality of the output can be expressed in measurable terms. Measurements can be used to inject feedback into the system to keep the process in control. This text will document and describe the controlled new product development process in complete detail. The process that is described takes advantage of the latest in tool technology. An integration and balance of best practices and tools are essential ingredients of the process.

New product development is a global process with global implications. Markets need to be viewed from an international perspective. Product definitions need to be sensitive to the needs of consumers around the world. Moreover, it's quite likely that the product's design will be able to take advantage of materials and processes that are available from other parts of the world. Manufacturing needs to adopt the best practices from leading-edge companies to remain competitive, which implies watching such companies. All functions within an organization must have the world perspective to maintain a competitive advantage. This parameter must be done by the fusion of observations made by an entire organization; it usually cannot be effectively accomplished by an individual.

Assembling the right team and having that team work effectively together form the essence of steps three and four. On the surface, it's a simple concept. For some, it is extremely difficult to implement. In future sections, particular attention is paid to defining what specific work needs to be done to develop new products and provide means by which the quality of the work produced can be measured. Knowing this information should simplify the task of deciding who is capable of performing the work and making effective assignments. Methods will be introduced that encourage team participation throughout the product development cycle. The group assembled to define the product being developed will contribute to its creation as the process involves. Nowhere in the process are abrupt personnel transitions recommended or encouraged.

Rapid prototyping, incremental product development, process models, electronic mail, and video presentations are a few techniques that can be used to enhance product and process visibility. Using these tools greatly simplifies the challenge of maintaining effective communications within the entire work environment. Keeping objects in view ensures that information about the new product development effort is constantly under active scrutiny. When minds are at work challenging other minds through visual experiences, the entire effort benefits. Throughout all remaining chapters, techniques and examples are offered to improve process visualization through the definition, planning, and execution phases.

Making effective decisions quickly is vital to the success of the new product development process. It's also paramount that just about everyone contributing to the process knows how to make swift, accurate decisions or knows how to get them made. Decisions must be synthesized vertically through hierarchical management layers and horizontally across functions. Obviously, communication links need to be immediate and groups need to be held down in size. Technique and adaptation of technique within a particular environment form the base for decision-making success.

Competitive new product development methodologies and tools become integral to the achievement of business objectives through the development of new products. The methodology and tools enable the new product development effort to move at a rate that is competitive. The new product development methodology offered in this text has been synthesized and refined over a period of years. In and of itself, it will probably function with only slight adaptation in most any new product development environment; however, it is not intended to provide process detail for outright utilization and implementation. The main thrust is to freeze a process description that serves as a reference for comparing other processes and ideas. By taking this action, it then becomes possible to make relative evaluations and judgments. The impact of new ideas can be evaluated. Strengths and weaknesses will appear; necessary adjustments will be made. From this experience, significant improvements to the overall process will materialize to make something that works extremely well in the new environment. Remember, the number-one action is to make new product development a controlled process; methodology and tools are significant contributors to the realization of the objective.

Attempting to create a new product with a pool of talent that is inadequately prepared, or lacks the skills needed to apply the knowledge they possess effectively, is an exercise in futility. Participants must be adaptable, creative, intuitive, knowledgeable, self-motivated—they have to be competent. An organization needs to know how to hire the right people with the right skills to do new product development. Recruiting quality college and experienced professional talent becomes critical to the survival of the organization. Upon becoming a member of the team, assignments must be appropriately made and trust between co-workers must develop—teamwork becomes critical. There is little opportunity to provide significant amounts of training. What training is offered comes fast and in large doses; the people at the receiving end must be able to rapidly assimilate information and have the skills necessary to effectively utilize what they learn. The new product development process works only if it is staffed with people who work effectively when they are told what to do and need minimal information and support on how to get the work accomplished.

Identification of process problems such as taking too long, spending too much money, and having things not work right is usually a lot easier than knowing where to focus attention to develop a solution. Because of depth in certain areas and breadth in others, it is often difficult to identify the source of problem production in the new product development process. Resources and time are not available to fix what isn't broken or attempt to fix suspect areas using the "shotgun" approach. Just as in any other process, the new product development process must be measured at frequent intervals. Measured data is analyzed to extract early warnings of potential obstacles. The acquired information is used to focus scarce resources on problem-producing areas to develop timely solutions. Measurement, data, analysis, and feedback must be relied upon to adjust the process—nothing else works!

Can do?

1.6 BRIEF PROCESS DESCRIPTION

Transforming an idea into a viable product in today's internationally competitive environment is a job that requires extremely talented teams of people who are well led and well managed. Just about everywhere you look there is information on being more competitive and productive. In fact, this information is so pervasive and voluminous that it is difficult to know where to begin its application. Those of you working on the development of new products want to apply the readily available information and tools to your work and many of you are doing so without the benefits of a focused vision on the overall process.

It is interesting to observe the makeup of the organizations and teams that do new product development. Many of them did not exist a few short years ago. Those that offer some degree of stability sometimes suffer from a high turnover of people. Success and growth can add people to the organization at a pace that exceeds the organization's ability to cope with the situation rationally. A decline in business can serve to eradicate entire departments. In almost all cases, inexperience in dealing with the problems at hand is in abundance, whereas experience is in short supply. Add to this the fact that change can further serve to diminish the small base of experience that is available and the uncertainties present in the working environment start to become self-evident. In short, things can be rather chaotic!

This text is designed to provide a framework and structure through time that can be used by organizations to become skilled at bringing new products to market. For a given new product development effort, time is divided into three segments. These are *definition, planning,* and *execution.* For each time segment, suggestions (based on data, test cases, study, and experience) are offered to enable each member of a project team to focus on the output they need to produce. The managers and members of the new product development team will be given tools and techniques to help them use the methodology. Output from each work group by process phase is specified using a matrix. The matrix model is displayed in Figure 1.11.

Examination of the figure shows that the matrix is constructed by arranging work function along the x axis and time along the y axis. Within each cell of the matrix, it then becomes possible to identify the work that must be accomplished by each function for each time segment of the new product development life cycle. Once the necessary elements of each cell have been defined, it becomes possible to arrange the information in each cell to produce a network of executable items. Where necessary, key nodes from an individual cell are linked to other cells. This horizontal and vertical linking of nodes within and between matrix cells produces a complete network model of the entire new product development effort. Thus, the matrix representation transforms itself to become a combined matrix and network model. From a modeling perspective, this structure may seem a bit unusual. The primary reason for using this hybrid matrix and network structure is to provide a simple framework that can be used to describe the new product development process accurately.

Future chapters will focus on the output from each of the phases and the means by which the output is produced. From a time-scale perspective, the definition phase could extend over a few months to over a few years depending on the complexities involved in getting technology and market into alignment. The planning phase can be done in a matter of a few weeks to a few months depending on the complication of the definition. The duration of the execution phase should be the approximate

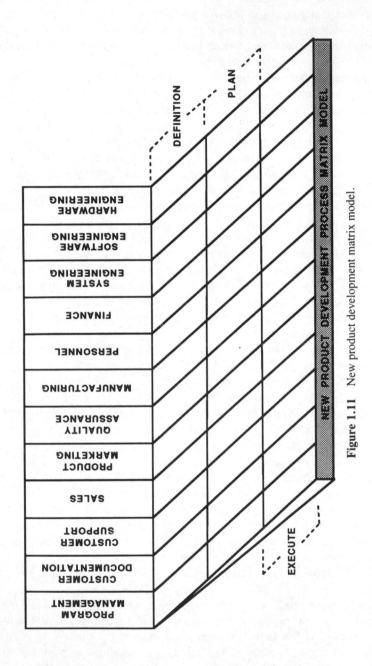

Figure 1.11   New product development matrix model.

sum of the durations of the definition and planning phases. Clearly, the organization has some control over the amount of time devoted to each phase and appropriate adjustments are encouraged. Work output from the definition phase drives the planning phase. During the planning phase, the execution phase network is synthesized. The network produced for the execution phase provides the structure necessary to complete the remainder of the new product development effort.

By utilizing appropriate elements from the information that follows, an organization will have the optimum probability of meeting its new product development objectives.

2

BUSINESS DEFINITION

In this chapter, you will learn about

The definition matrix.

The creation of a vision for the business.

The generation of a "big picture" for the business.

The business application of market information.

The synthesis of business alternatives.

The analysis of business alternatives.

The creation of business definitions.

A slow strategy is as ineffective as the wrong strategy.

Kathleen M. Eisenhardt

2.1 BUSINESS DEFINITION SYNTHESIS

In this and the following chapter, the structure and the details of the definition matrix shown in Figure 2.1 will be explained. The business column is completed first in the vertical cell sequence given. At certain times, there are "loops" between cells. Most often, iteration occurs between the analysis and synthesis cells. Output from the business column becomes the initiator of product column activities. This chapter is dedicated to the business column. The next chapter will focus on the product column.

To make money, it is necessary to deliver something that the holder of the currency wants. The definition matrix helps make this simple concept possible. A modern business exists to provide products and services in exchange for customer payback. If the business is operating with some degree of success, the enterprise will return a profit. Profit dollars are reinvested into the business to sustain financial momentum. More often than not, a percentage of profit dollars is invested in new products and services to maintain and grow the business. Product and service definition decisions made today will determine the future success of the business. To achieve success with new products, it is essential to think about definition synthesis from two perspectives: the business definition and the product definition. The former can lead quite nicely to the latter. Jumping directly to product definition without a business definition can lead to financial and business misfortune. This section deals with the generation of a quality business definition that will lay the foundation for the new product development process.

	BUSINESS	PRODUCT
VISION	VISION OF THE BUSINESS CORPORATE OBJECTIVES 3 TO 5 YEAR OBJECTIVES	INTERNAL PRODUCT PURPOSE EXTERNAL PRODUCT PURPOSE
BIG PICTURE	NECESSARY PRODUCTS NECESSARY SERVICES TARGET CUSTOMERS CHANNELS OF DISTRIBUTION COMPETITIVE PRESSURES	PRODUCT DESCRIPTION TARGET CUSTOMERS COMPETITIVE POSITION
MARKET INFORMATION	CUSTOMER DATABASE PRODUCT DATABASE COMPETITOR DATABASE MARKET DATABASE DATA REDUCTION	NECESSARY PERFORMANCE NECESSARY FEATURES NECESSARY COST NECESSARY PRICE NECESSARY TIMING
ALTERNATIVE SYNTHESIS	DESCRIPTIONS & SPECIFICATIONS TECHNICAL APPROACH TECHNICAL RISK INVESTMENT TIME TO MARKET	INTERFACE DESCRIPTION SPECIFICATIONS TECHNICAL RISK INVESTMENT AND TIMING FIT WITH PRESENT SKILLS
ALTERNATIVE ANALYSIS	IMPACT ON THE MARKET RETURNS FROM THE MARKET FINANCIAL ANALYSIS POTENTIAL PROBLEMS	IMPACT ON THE MARKET RETURNS GENERATED DETAILED FINANCIAL ANALYSIS POTENTIAL PROBLEMS
OUTPUT	CORPORATE BUSINESS DEFINITION DIVISION BUSINESS DEFINITION PRELIMINARY PRODUCT DEFINITIONS STRATEGIC RECOMMENDATIONS	PRODUCT DEFINITION NPDP MATRIX ENTRIES COMPLETE RECORDED ESTIMATES

(Row group label along left margin: **D E F I N I T I O N   M A T R I X**)

Figure 2.1 Definition matrix.

The goal of this chapter is to provide sufficient detail to allow you to participate in the formation of a business definition for your organization. For those of you fortunate enough to be working in an organization where a business definition is in place, you can use the material to evaluate your present situation. Ideally, each and every business definition should contain key information from each cell within the business column in Figure 2.1. Having the requisite business structure firmly cast makes it possible to create quality, stable, new product definitions that are fully understood by horizontal and vertical new product development participants. Under certain conditions, the work needed to satisfy all cells in the business column has been done in a day. Other circumstances may dictate a longer time commitment. Once this learning experience takes place, it then becomes possible to make adjustments and improvements.

A business definition consists of six components: a "vision," a "big picture" description, market information, a synthesis of product alternatives, an analysis of product alternatives, and output from the business definition process itself. Each element is now considered in greater detail. The discussion will follow the order of the business column of the definition matrix.

2.1.1 Business Vision

A business vision is intended to empower, motivate, and focus the work force within an organization. It's a relatively easy exercise to determine if your organization has a vision, but how do you go about determining if the available vision is a good one? If you think it's weak, where do you make change for improvement? Should you be attempting to create a business vision, how do you know when the effort is complete? In this section, questions of this type will be answered while the contents of the definition matrix cell are fully explained.

The following criteria that can be used to evaluate the quality of a business vision have been assembled based on the analysis of many "vision" statements published by corporations or made by their leaders:

1. Everyone within the organization, from line worker to executive, is aware of and understands the statement.
2. The statement is short, bold, and optimistic.
3. The statement is repeatedly made by the right person, a leader in a position of power within the organization.
4. Creative thinking is stimulated by the statement.
5. No mention is made of products and services in the statement.
6. The statement places constraints on the organization's market focus.
7. The statement looks well into the future; its views go beyond next quarter's earnings or this month's production schedule.

For discussion purposes, let's look at a vision statement extracted from a corporate brochure:

Lockheed is a worldwide systems-engineering and high-technology leader. Our mission is to meet the needs of our United States and Foreign customers with high quality products and services, and in so doing produce superior returns for our shareholders and foster growth and achievement for our employees.

—Lockheed Corporation

Lockheed, a highly respected company, is not alone in producing limited message vision statements. There are more vision statements of this type coming out of major corporations than can be counted. More than likely, the company you work for has something like it. How does it stack up on the seven points mentioned above?

Are people within the organization aware of and do they understand the statement?

Answer: no.

What do terms like *systems engineering, high-technology leader, United States and Foreign customers, high quality products and services,* and *growth and achievement* mean to the average Lockheed employee coming into contact with this statement? The only way for words of the type selected within this statement to make sense is to footnote them and then go about supplying a definition in the footnote! Obviously, a statement that cannot be understood is not likely to be remembered. A sampling of Lockheed employees were asked if they knew what their company's vision or mission was. So far, without exception, no one contacted for the sample knew the answer.

Is the statement short, bold, and optimistic?

Answer: no.

It's none of the three. The statement contains almost 50 words. This fact could be overlooked if the choice of words conveyed a sense of boldness and optimism to the reader; unfortunately, they do not. This is the kind of statement that makes the mind wander to other more pleasant thoughts.

Is the statement made by the right person?

Answer: no.

An individual would not make a statement of this type. This statement was drafted by a committee! Committees distribute such material and think they will be assimilated by the global population. When was the last time you were inspired by the vision of a committee?

Does the statement stimulate creative thinking?

Answer: no.

Maybe a direct "no" is too harsh. A number of people who came into contact with Lockheed's vision statement said that it really got their creative juice flowing. Of course, when asked where they were taking this creative opportunity, the answer was that they were rewriting the vision statement!

Are products and services omitted from the statement?
Answer: no.

Not only are they mentioned, but they are specified to be of high quality. Product and service offerings are implied in business. Why state the obvious?

On a positive note that was probably the result of a group accident, no specific products and services are mentioned. This is a highly desirable attribute.

Are constraints placed on market focus?
Answer: no.

This is a statement of do all for all. Clearly, no viable business enterprise can operate in this manner for a sustained period of time.

Does the statement look well into the future?
Answer: no.

The statement is so vague and dull that it makes the receiver of its message tired. A "turned off" individual is not likely to apply much energy in helping create a better tomorrow for the company and its customers.

In summary, the vision statement just examined fails on all seven evaluation criterion. It is difficult to comprehend how a company can publish a vision statement of such questionable value. What did the individuals who drafted the statement expect it to accomplish? Maybe there is a connection between the quality of a corporation's vision and its business success. Right now, Lockheed appears to be headed for hard times as the deep pockets of its government customers empty.

In contrast to the previous vision statement, let's examine the following:

Our mission is to make the MIPS architecture pervasive worldwide.
—MIPS Corporation

Are people within the organization aware of and do they understand the statement?
Answer: yes.

If you ever have the opportunity to talk with a person from MIPS, ask them what their company's vision is. I do and I always get the same right answer! This statement is easy to understand. In fact, the only possible point that needs clarification is to try and determine just what is the MIPS architecture.

Is the statement short, bold, and optimistic?
Answer: yes.

The statement consists of 10 very powerful words for a company the size of MIPS. To have their architecture become pervasive on a worldwide basis is bold and optimistic in view of the competition in this market.

Is the statement made by the right person?
The statement is frequently made by Robert Miller, the chief executive officer of MIPS.

Does the statement stimulate creative thinking?
Answer: yes.

The vision is open-ended from strategic, tactical, and logistical points of view. It encourages the questions, Just what does the MIPS architecture have to be to be pervasive? How do we, as employees, make the architecture pervasive on a worldwide basis? Keeping just these two questions answered over time can grow and develop the creative spirit in MIPS for many years.

Are products and services omitted from the statement?
Answer: yes.

An architecture is an invisible artifact. It is a skeleton upon which a business can be built. Business success is obtained by the transformation of the architecture into products and services that meet the needs of the market over time. This statement starts the product developers at MIPS thinking about what their product offering should be. It will lead to the creation of a product and service strategy that will be implemented over time by the MIPS organization.

Are constraints placed on market focus?
Answer: yes.

The market is the worldwide market where the MIPS architecture can effectively perform. This is an appropriate constraint.

Does the statement look well into the future?
Answer: yes.

This statement challenges the entire MIPS organization. It's motivational; it gives every employee something to strive for. The company's current position indicates that it has a significant hurdle to jump to have its architecture become pervasive on a worldwide basis.
It's clear that the MIPS organization has a business vision that is very high in

quality. Also, as a business entity, its performance has been rather impressive in recent years. It would be difficult to say that its vision has had no impact on the organization's performance.

Having reviewed example visions at opposite ends of the spectrum, you are now in an excellent position to perform an analysis on your company's vision. The obvious, but sometimes very difficult, first step is to locate the vision statement. Once this has been accomplished, you can then answer the question, How well does it perform on each of the seven evaluation criteria? In areas where weakness has been detected, what can be done to make improvements? Keeping in mind that improvements in this cell of the definition matrix can only come from the top management of the organization, it is highly recommended that top management take hold of the situation and work to make the necessary revisions. People at lower levels can help with analysis and provide meaningful feedback, but only those in control of the entire organization can prepare and then communicate the "vision." Where possible, the organization's number-one person should take full responsibility for this extremely important work. All new product development participants are in an excellent position to bring problems with the "vision" to the attention of top management, but only top management is in the position to act.

Once an effective vision is in place for the organization, attention can shift to reviewing the guiding objectives of the entire organization. The objectives most frequently cited by established businesses tend to expand on such subjects as profitability, growth, fields of interest, customers, people, management philosophy, and citizenship: It just makes good business sense to make money today, to make more money than today tomorrow, to focus on where to receive compensation, to receive compensation from a satisfied group, to produce products and services in a humane fashion, and to return a portion of success to the surrounding environment that helped make it possible. Rather than further discuss each of these topics, an example "statement of objectives" is included in the Appendix (see "Example Corporate Objectives" therein) to serve as a guide and reference for those of you seeking to document your company's broad objectives thoroughly.

For a different perspective on corporate objectives, let's examine the five primary objectives of the Disney Corporation:

Think tomorrow.
Free the imagination.
Strive for lasting quality.
Have "stick-to-it-ivity."
Have fun.

On the surface, this view appears to be entirely different from the conventional format.

In the Disney sense, "thinking tomorrow" suggests doing today's work so that you can get a payoff tomorrow. It's an attitude. This attitude in its literal interpretation sets the expectation that the company will receive compensation for the work it has performed to provide products and services. There is also an alternative interpretation. "Thinking tomorrow" gives the organization the flexibility to examine today's mistakes, to weather disappointment, and to ultimately return with something so powerful that its success is guaranteed. These two small words are capable of producing significant business rewards. Certain business situations demand a good night's rest before progress can be made. An organization that understands this basic principle and encourages its employees to briefly walk away from problems sets itself up for achievement over the long term.

An organization with a free imagination can readily identify opportunity once evidence of it begins to surface. Seeing, touching, hearing, and just plain old thinking can often be enough of a stimulation to launch a major undertaking. No one knows where the inspiration will come from, but an organization comprised of people with free imaginations working to make something happen is certain to seize opportunities and capitalize on them the minute their existence is suspected. This objective encourages inspired risk taking. It produces an environment in which people will usually accomplish far more far better than they thought possible.

Disney has been successful at turning out products and concepts of lasting quality. Mickey Mouse is now over 50 years old. The first Disney theme park opened over 35 years ago. Movies like *Snow White and the Seven Dwarfs, Pinnochio, Alice in Wonderland, Peter Pan,* and *Twenty Thousand Leagues Under the Sea* continue to excite audiences around the world and defy the effects of time. Achievements of this magnitude don't happen by accident. It's clear that lasting quality is an objective that Disney tries to achieve with every product.

According to John Culhane, "stick-to-it-ivity" is a word invented by Walt Disney that was sung by an animated owl in the 1949 movie *So Dear to My Heart.* It means to hang on to visions, values, and beliefs even when conditions look dismal or if the going is demanding. Everyone experiences hard times. Knowing how to persevere and survive less than ideal situations builds character and builds staying power in business.

Having fun is virtually a by-product of the first four objectives. I'm entertained and amused when I think about how I can use tomorrow to finish what I started or to correct the mistakes I've made today. You'll often find me in a playful mood when my imagination is working overtime on an exciting concept. I really enjoy seeing the work I do stay around for a while; after all, why would I even consider writing a book? Surviving the lows of business really turn the highs into moments of extreme pleasure. Yes, for me, the Disney objectives are what fun is all about. How do you see it?

The Disney objectives reside at a higher level in the objective hierarchy than profitability, growth, fields of interest, customers, people, management philosophy, and citizenship. Achievement of the Disney objectives depends on satisfactory accomplishment of the lower-level objectives.

Hopefully, the detail supplied in the Appendix and the concepts provided in previous paragraphs will be sufficient to make the analysis and synthesis of objectives for your corporation a well-guided activity. Again, analysis of existing objectives and the synthesis of new ones are work reserved for the top management of the organization. Members of the new product development teams may contribute, but only those in control of the organization can make the final determination of the conceptual level and the content they want to communicate with their objectives.

With the organization's vision and guiding objectives in place, it becomes possible to utilize the material to produce objectives that can be accomplished over a particular period of time. Frequently, those objectives with a time orientation are referred to as two- to five-year objectives. Most often, these objectives will speak to gross revenue, profitability, expenses, market share, and other business specific items. Example statements follow:

- Maintain worldwide net profits at greater than 8 percent of sales.
- Increase gross sales to greater than 15 percent per year.
- Increase expenses to no more than 10 percent per year.
- Form alliances with leadership companies with similar business interests.
- Become a $50 million company in five years.
- Increase market share at greater than 5 percent per year.
- Develop at least two new major markets.

The above objectives become points of reference at the corporate level that are communicated to operating entities within the corporation. Operating entities use this information in setting their objectives from a top-down guided perspective. As entity information is returned, a bottom-up integration process can be followed to arrive at a final summation for the overall corporation.

As you have read, the vision for the corporation starts at the top. It consists of three elements: (1) the vision of the business; (2) the broad corporate objectives; and (3) two- to five-year objectives. All elements are provided by the top management of the corporation. Feedback from operating entities is analyzed and adjustments are made to arrive at a final comprehensive "vision" for the corporation. Doing what is contained within the "vision" cell starts the process of management by objective. You'll also find this to be the conceptual starting point for the production of Hoshin-style decomposition of objectives and goals throughout a business entity.

To help clarify concepts contained in the business column of the definition matrix, examples of corporate and division business definition documents are presented in the Appendix. These documents present relevant information through a company called MONI. MONI is a fictional extension of public information available on Sony Corporation. The Sony data have been used to create the documents that illustrate how the information being discussed within the definition matrix can be applied. In this way, information that a real company would consider confidential can be made available to clearly illustrate the information content and concepts behind the creation of business and product definitions.

To observe how MONI implemented the concepts discussed in this cell of the definition matrix at corporate and division levels, please refer to the Appendix. Examination of the corporate and division business definitions shows how the display division made use of the corporate vision and the five-year objectives to set its own purpose and objectives.

2.1.2 Business Big Picture

Emphasis now shifts to products and services. Here, top management has the task of identifying the necessary products and services it wants the corporation to produce, distribute, and support. Attention must be given to the identification of target customers while an eye is kept on competitors with similar interests.

To become familiar with "big picture" concepts, imagine yourself as a top executive of a major corporation. If you like, think of yourself as the CEO! As a member of this select group at the helm of a multibillion-dollar enterprise, you are responsible for defining the necessary products you want your company to produce, the necessary services your company provides, the customers you will target, your primary channels of distribution, and your competitive philosophy. Information you assemble about these topics will permeate your company. As you might expect, this broad detail will guide the behavior of your entire organization; it can be expected to be highly leveraged. The production of output within this cell starts with being able to identify target customers.

For starting convenience, establish the primary geographic regions where you would like to do business. Is it your intention to have your company become a local, regional, national, or international provider? Once you have a geographic focus, it becomes possible to develop a profile on the people you want to service in particular areas. A medical electronics company may have its sights on men and women with a specific type of illness. A software company may want to target individuals working in a particular industry with a particular set of problems. You'll next want to give consideration to other human factors such as sex, age, profession, ancestry, education status, and economic status. Focusing your corporation on a specific segment of the population can sometimes be an important business advantage. Think through a number of situations that differ from yours to refine your technique. How would a tobacco company describe its target customer? Do you think Harley-Davidson has a profile on the type of person who is likely to buy a "hog?" Lastly, give thought to how you want to serve your customer over time. Many companies operate with a cradle-to-grave view of their customer. Sony would like to initially service your children with "Their First Walkman" and your money! Once the die is cast, Sony provides a continuous flow of products and services over the individual's lifetime. Auto companies try to do the same by having economy brands to attract the young, first-time buyer while providing up-market alternatives as loyal customers mature. Other providers make their contribution at particular times in their customer's life. Our education system is probably the best example of a business that is tightly segmented by customer age. Where on the human time line do you want to fit? As your collective view of your target customer materializes, you then are in a position to better understand his/her problems, needs, and wants to focus your company on your customer effectively.

With knowledge of your company's customer reasonably well refined, you can then give serious thought to products and services you would like to provide. The technique that is recommended trying first is a relatively simple one. Obviously, you already have some idea about what your product and service offering is, or is about to become. Put this knowledge to good use by compiling a list of what you are doing or intend to be doing. Those of you fortunate enough to have a short list can then create an inspired statement about the limited number of products you have to consider. A one-product-line company like QMS Inc. has a relatively easy job of this. It simply has to say a few kind words about its focus on monochrome and color printers. For example, QMS had this to assert in their 1990 annual report:

> QMS Inc. provides its customers throughout the world with innovative computer printing systems and solutions that are recognized as among the best in the world, both by customers and experts in the field.

Note the simplicity in the statement. After reading it, do you have any doubt about the product focus and geographic scope of QMS? If your listing of product lines grows to be more than a few items, you'll have to attempt to create a hierarchical decomposition of products produced by your company and try to create a sensible grouping.

To demonstrate how this is done, let's examine a sampling of the product offering of Sony:

"Walkmen"	AM/FM headsets
Portable CD players	Cordless telephones
Portable DAT players	Shortwave radios
Receivers	Speakers
Amplifiers	Audiotapes
Videotapes	CDs
Disk drives	Theatrical productions
Theme parks	VHS camcorders
8-mm camcorders	"Watchman" TV
Still video camcord-	Portable AM/FM/tape
ers	
Portable AM/FM/TV	Portable AM/FM/
	tape/CD
CD players	Tape recorders
Displays	TV receivers
VCR	Laser disk
Home theater	CD ROM

Do you see a way to segment Sony's product offerings? All that needs to be done is identify product categories and assign products to the category that appear to offer the best fit. One possible alternative produces eight product classifications. The primary decomposition and products that are subsets of the exercise follow:

Personal Audio Products	**Home Audio Products**
AM/FM/CD players	Tape recorders
"Walkmen"	Receivers
Portable CD players	Speakers
Portable DAT players	Amplifiers
Cordless telephones	
Shortwave radios	

Personal Video Products	**Home Video Products**
8-mm camcorders	TV receivers
VHS camcorders	VCR
Still video camcorders	Laser disk
"Watchman" TV	

Personal System Products	**Home System Products**
Portable AM/FM/tape	Home theater
Portable AM/FM/TV	Integrated audio/video
Portable AM/FM/tape/CD	

Software Products	**Computer Products**
Audiotapes	CD ROM
CDs	Displays
Videotapes	Disk drives
Theatrical productions	
Theme parks	

With this decomposition in place, the Sony management team can go about deciding how to build an organization to manage the eight classifications of products. Should it be four primary groups that consist of personal products, home products, software products, and computer products? The personal products group would have responsibility for personal audio, video, and systems products, whereas the home products group would manage home audio, video, and system products. Maybe it would be better to have an individual division work on each of the classifications? This scenario would yield a corporate hub with eight divisions. Quite possibly, the company should organize around audio, video, and systems. These are not easy decisions and any decision made at one point in time is subject to revision as a function of business dynamics. Nonetheless, the top management team at Sony has to work on issues of this type. Once decisions are made, they must be implemented and supported. Try to imagine the decision process behind creating the organizational infrastructure to deal with a product mix as diverse as Sony's!

Few of you working on new product development will have direct responsibility for dealing with issues of this type, but it is extremely important for you to know who is supposed to be doing this work. When top management does its job correctly, you will fully appreciate the decisions that were made and the infrastructure that results. When management's performance is weak, you may find yourself in a new product development environment that may be intolerable. Sony is an example of a company that appears to be organized in an extremely effective fashion. Many organizations wish they could do as well! You need effective product-line market focus and a somewhat stable organization before you can do meaningful work on new products.

With necessary products and services defined, the company can now give consideration to distribution issues.

A distribution channel is in place to perform all the activities needed to move a product from production to the customer. First, the product needs to be sold; then, it must be delivered. Sales may be generated by indirect and direct means. Indirect sales methods such as direct mail, telemarketing, radio and TV advertising, and so on, are effective for certain product types and weak for others. Direct sales through a dedicated sales force, sales representatives, or retailers offer other alternatives with unique challenges and opportunities. Product delivery follows the sale. Certain products may flow directly from factory to consumer. Being able to operate in this mode is usually accomplished by taking orders from a network of sellers and order-processing stations and making product delivery to the customer within a tolerable period of time. Where customers can accept a day or two delay in receipt of product, a manufacturer can make delivery using the services of companies like United Parcel Service (UPS) and Federal Express. Other situations may require that products be transported to warehouses or to local showrooms. When this takes place, distribution enters the domain of wholesale and retail networks and systems. Large companies sometimes find it necessary to create their own distribution network of warehouses, sales offices, and delivery systems. Other firms may find it more attractive to work through an existing distribution system. Of importance to the new product development team is knowing what distribution systems are available, which systems are preferred by top management, and guidelines on distribution system utilization.

I am associated with a company that has an elegant distribution strategy. Its president provided the following guideline: "If the product can't be sold through magazine advertising and delivered by UPS, don't do it!" Those working on the development of new products at this company have no doubt about how products will be sold and how they will be delivered. Do you think having this a priori knowledge influences new product design and process decisions? Of course it does! The design of the product and its shipping container are done in concert to ensure it can withstand the worldwide shipping rigors of planes, trains, and trucks. Shipping containers are not made to be pretty, but to work. After all, the shipping containers are not going to be seen on any retail shelves. Product design and product support are influenced by this constraint. Designers know when the customer receives the product there won't be anyone around to help them use it. They therefore design the product to make using it right out of the box easy to do. Features to facilitate setup, turn-on, test, and fault diagnosis are designed into the product. Documentation is designed to complement the product's internal design. Advertising, sales literature, and user documentation work in harmony to make the strategy work. In the unlikely event of a product malfunction, an empty return shipping container is supplied free of charge just be making a phone call or sending a fax. New product development team members appreciate having a top corporate manager provide this information.

Firms operating with the simplicity of distribution previously described are in the minority. Most often, a wide number of alternatives will have to be considered. From the evaluation, a few will be selected. Only those corporations with enormous wealth or those that are foolish, or both, will attempt to develop all available distribution channels for their company. Top management is in the position to influence and manage the distribution channel using new products as a tool. Communication of intentions is the first step in the process.

Many businesses compete with a broad product base of competitors. For example, in consumer electronics, RCA, Mitsubishi, Panasonic, Canon, Ricoh, Chinon, Pioneer, Japan Victor Corporation, General Electric, Kyocera, Toshiba, Magnavox, Zenith, Technics, NEC, Sharp, Proton, Sansui, Yamaha, Sony, and Philips have the potential to compete against each other on a number of different fronts. Most often, a specific product line will have a substantially reduced number of competitors. Top management should not be tasked with providing specific information about each competitor; rather, they should be held accountable for setting the tone on how they want their company to compete. The determination and communication of values held by the top management team help influence a company's competitive behavior. A firm that has no difficulty taking like copies of competitor's products to market is likely to compete differently than a firm that finds duplication abhorrent to their very being. A firm that thrives on marketing warfare ideals contrasts with a competitor that elects to succeed in business by forming close relationships with customers. The new product development process will function with greater effectiveness once top management effectively establishes the organization's competitive position.

In summary, the elements making up this big picture are (1) a list of necessary products and services; (2) the identification of target customers; (3) an overview of distribution channels; and (4) a brief discussion on competitive pressures. Top managers of the corporation are tasked with providing quality information on each of the topics. Refer to the Appendix ("Corporate Business Definition Example" and "Division Business Definition Example") to see the output produced by managers of MONI Corporation at corporate and division levels.

Knowing the elements of the big picture cell is generally sufficient to build this section of the definition matrix for your company. Hopefully, the top management team will take the initiative to provide the information requested within this cell with minimal pressure from new product development team participants. With just a little effort, the information that is needed is typically easy to obtain and the data requirements are minimal. A small, top-level management group assembled for a day or two can generally produce a high-quality big-picture draft. The draft should be distributed to general and functional managers for their review and feedback. Once all of the reviewers comments have been collected, it is then a simple matter to make the final adjustments and complete the work on this area.

2.1.3   Business Market Information

The output from this cell of the definition matrix is produced by the top management team of the corporation. Reducing information about products, customers, markets, and competitors, the management team produces "guidance" statements for all new product development activities within their organization. This output is not focused on individual product efforts; it is intended to encompass global issues that affect product lines as a whole. Information from this cell establishes management's personal and financial commitment to particular areas engaged in new product development activities. The following discussion explores a possible path for producing the needed information. Make appropriate departures to suit unique business situations.

Market models derived from a business database can play an important role in helping determine the upside potential for new product or service ideas. With suitable models, it becomes possible to answer questions about the market's overall size and total growth rate, how the market segments, how the product or service idea might impact the market, how each segment is performing, who key customers are in a segment, how market share is distributed between competitors, and so forth. As a business entity embarks on collecting data for each field contained within the database, it becomes clear that the volume of information to manage has the potential to grow to a very large number of elements. A direct, numbers-only, approach to construct a market database usually leads to collecting, adjusting, and maintaining a vast amount of market data. Frequently, assumptions need to be made to simplify the data collection and manipulation task. Intelligent data reduction and interpretation are usually performed by skilled human interaction with the data and the database systems. Once a balance is achieved between numerical and intellectual methods, it is possible to construct a reasonably good database for the market that can be refreshed with current information. A complete and accurate database provides decision makers with a common view of the market to enable them to make better new product development decisions for their customers.

Computer systems enable amassing much of the information required by a sophisticated market model. Systems are capable of tracking a serialized product from the shipping dock directly to any sales outlet in the world. A system of this type makes it possible to fully characterize product distribution as a function of international geography. Corporate executives, with the desire, can determine who sold each product, its price to the seller, inventory levels at the seller, and the time the unit remained in inventory. Collecting this data over a period of time and channeling it through data-processing algorithms allow measurement of other important parameters. Sales growth by product type by geographical area is produced by comparing the data over two different periods of time. Seller effectiveness can be evaluated by measuring the sales for a given operation over a period of time. Locating where certain products sell best is simply a matter of examining the data by sales region. Effects of regional promotional campaigns can be evaluated by measuring sales rates before and after the programs. The creative mind can manipulate the information from this system and use it to enhance measurement and modeling performance. For all its benefits, however, this tracking system yields almost no information about the customer.

To reach the customer, many companies rely on the individual taking possession of the contents of the shipping container to complete and return the warranty card attached to the product. At times, the customer is told that the product must be registered for the product's warranty to be valid. Statistically, systems of this type can be structured to provide very useful information. A corporation using this technique audits a percentage of the returns to ensure their statistical accuracy. This provides a company with important customer data. Required information consists of name, address, retailer name, and date of purchase. Supplemental information varies from product to product, but typically consists of questions about the age, marital status, and sex of the purchaser, profession, employer; the annual salary of the purchaser; the reason for the purchase; the product's application; the reason for selecting the product over others; the names and ages of other users; the other products produced by the company that the customer owns; and a phone number that can be used to obtain added information. Thus, a corporation is able to collect adequate information to better understand the *purchaser* of the product, the *user* of the product, and the intended *application* of the product.

Amazingly, present technology can process the millions of pieces of information contained within a database into an understandable form. This is a first step in gaining some analytic representation of the business environment. Business managers now have the capability to view product sales as a function of geography, product type, and customer classification. Snapshots of the data can be summarized on a daily, weekly, monthly, or quarterly basis and analyzed to extract trending information. Clearly, intelligent use of this information can be of extreme business value, but something is still missing. What about the competition?

A company can usually find a way to do an excellent job of collecting data about the products it sells and the customers who buy its products. In market segments where the company is the market leader, this might be an acceptable situation, but in segments where a company is one of many players, getting some information on competitors' performance is of vital necessity. How else can the corporation go about calculating its market share? Why have certain individuals decided to make a purchase from a competitor? A company must now devise some clever means of estimating the share of the market held by each of its competitors and develop an understanding of why it cannot win with everybody. From these estimates, it is then possible to predict total market size and compute the share of the market held by each participant while determining reasons for loss of business to the competition. This is no easy task; and, to make this a meaningful exercise, it must be done fast enough so that the data collected is not out of date before it's used to make a key decision. This means that whatever process is followed to collect the information, it must be automated to keep the database from rapidly losing its significance.

Some corporate entities maintain a staff of market research specialists whose primary job is to collect competitive data and generate overall market penetration figures for their enterprise. These specialists are skilled at extracting data from public documents (quarterly and annual reports, credit reports, trade magazines, news releases, analyst recommendations, advertisements, etc.) and private sources. In effect, they scour the world for important market information. Frequently, bits and pieces of data need to be thoroughly analyzed and expertly reduced to extract the critical information that is locked within their confines. As the market research specialists gain confidence in the accuracy of the information collected and the techniques they use to reduce it, their contribution to the modeling effort begins to materialize. Business leaders now get quality information on the dollar size of each market they serve, the dollar size of key segments of each major market, the competition in each segment, and the market share held by the competitors. Updating this database over time gives the market research staff the ability to do trend analysis and develop algorithms to forecast market and competitive behavior.

Decision makers now have a powerful decision-making tool at their disposal. Their database is rich in product, customer, competitor, and market information. Creative use of the data can lead to the generation of extremely valuable decision models, graphical models, and mathematical models. Additionally, database sorts can be used to extract other pieces of key business information. Top corporate management is now equipped to use the data to make and in turn to communicate its interpretations and decisions. Within corporate executive offices, data reduction, analysis, and decision making leads to the production of statements like

To improve and speed decision making and bring erosion of market share under control, (your company) is merging four groups into two.

The market continues to grow largely due to the double-digit demand for our recently introduced products. (Your company) will continue to invest in this product line at present rates to increase market share.

Early sales results from our new venture are encouraging. Market size and growth potential suggest that we will continue to invest in this opportunity above normal levels.

The present commercial market is small but growing rapidly. Most sales were going into military applications. This is an area where the management team wants to continue investment in new product development with strong emphasis on commercial applications and deemphasis on military product development.

Sales on our products within the line are essentially flat or slightly declining. Look to move the manufacturing of the product line to areas that offer better competitive advantage. Funding for new product development will be reduced by at least 30 percent.

Our growth rate continues to outpace that of the market, a trend we want to see continue. Strive to make technology advances a reality. Development funding will grow slightly over last years levels.

Our attempts to make last year's big hits incrementally better have failed. We will research the market and determine probable cause. While this effort is taking place, new product development staffing and funding will be dramatically reduced.

Our participation in this market is facing stiff competition and heavy losses. Plan to exit from this market. There will be no funding allocated to new product development.

The management team has processed a vast amount of data to arrive at the production of "guidance" statements for new product developers. The preceding examples illustrate a management presentation style that is concise, direct, and understandable. Managers and new product development team participants now know where their efforts fit into the global corporate environment.

Top management's use of the database to produce guidance statements for new product development is just one of many potential applications. Operating units, departments, and individuals with database access can process information in the database to focus in on their particular area of interest. Use of the market database becomes a hierarchical process. Those who need more detail can extract it, those who do not are not submerged in a sea of excess data.

Every business that has a product available for sale has the potential for creating a market database. A new business has the potential for creating the database once new products are available for sale. Properly structured, the market database can yield vital information that can be reduced and communicated to the organization. Refer to the Appendix ("Corporate Business Definition Example") to review the results produced by MONI management after processing product, customer, competitive, and market information within their database. The section entitled "Product-Line Guidance" in the corporate-level MONI example is provided to illustrate one effective means of using the database to produce business information vital to the operating performance of new product developers. Pay close attention to how guidance at the corporate level of MONI was assimilated by the display division in its business definition document ("Division Business Definition Example" in Appendix). In future cells of the definition matrix, the role the market database can perform in helping establish quality product definitions will be explored.

2.1.4   Business Alternative Synthesis

Most organizations contain individuals, and sometimes small groups, that are skilled at generating ideas for the future projects the organization might undertake. These individuals tend to be highly visible and are usually heard from more than once if properly motivated. On the other hand, an environment that provides limited opportunity for presentation and consideration of new product ideas is likely to silence or drive away this type of person. These people exist within every organization. They can provide an abundant supply of ideas on what should be done next. The challenge to the organization is to encourage and then manage the flow of new product ideas without discouraging their production. A company should want to make it known that new product development ideas are welcome and also provide the developers of the ideas with a process to follow to get their ideas evaluated.

Teams, organizations, and very large groups don't have it within themselves to create new product ideas, but they do have the ability to critique, analyze, and develop the idea. Most of the time, an untrained organization will follow the order just given and more often than not stop at the critique phase. It's unusual for someone to say, "Great idea, let's do it!" This would require that everyone in the organization have the same authority as the president. Building an environment that links the creation of new product ideas to their collective development is an important facet of the business synthesis process. The challenge to an organization becomes one of selecting the best ideas while delaying all others. Remember, there is no such thing as a bad idea, but there are ideas whose timing, focus, and scope need continued adjustment. Selection of the "best" new product ideas has to be made based on a wide variety of considerations and with a fair number of people wanting to gain recognition and make contributions. Getting ideas into the open for their first objective review is the main object of business alternative synthesis. This is the first step in gaining organization support for the idea.

Depending on the type of organization, decisions to invest in an idea will be made by an individual (owner of the company) or a team that contains a dominant individual (president, general manager, etc.). Within this cell of the definition matrix, we'll develop a scenario that can be applied within a large corporation. This structure of the organization will contain the idea people, strategic business units, divisional general management, and corporate management. The roles performed by each group will be detailed as they deal with alternative synthesis issues. Smaller companies can simplify the structure provided to match their business situation.

Strategic business units (SBUs) in some companies are sometimes called business teams in others. They consist of middle managers from research and development, marketing, and manufacturing who are responsible for product-line management within a division or operation. The SBU is accountable for product-line strategy, tactics, and logistics. Of course, this team takes limited action without upper management authorization. The idea people within the business are informed that new product ideas get their first review at the SBU level. Additionally, the SBU requires idea submission to be accompanied by a minimum level of support information.

The system must be fair, equitable, and fast. No employee or contributor should be discouraged from participating. Similarly, consideration requirements should not present too high an obstacle to discourage presentation of ideas, nor should they be so simple that poorly thought out ideas are readily submitted. A two-tier evaluation approach often works best. The first tier ensures that ideas are within the scope of the company's and its division's visions and objectives. Ideas that successfully exit the first tier move to the alternative analysis cell of the definition matrix. First-level evaluation is performed by the SBU.

While not a requirement, it is a good idea to suggest a guided format for initial new product idea presentation. It's fair to request some information on

Visual images	A picture, sketch, on screen computer display, or video that adds some type of substance to the new product idea.
Company benefit	Brief comments on what the new product idea does for the company.
Customer benefit	Brief comments on what the product idea does for the user.
Description	A short written description of the product.
Customers	Preliminary thoughts on who would be attracted to the new product.
Competition	Identification of similar products on the market and the providers of those products.
Features	Identification of product attributes that should appeal to customers.
Specifications	Numerical performance parameters the new product is projected to produce.
Resources	Early thoughts on the number of research and development, marketing, and manufacturing people it will take to define, plan, and execute the development of the new product. Suggestions on resource timing should be included.
Sales	A suggested sales price; thoughts on anticipated mature sales; and preliminary time-sensitivity estimates that lead to the production of a market window value. If the product idea impacts present product sales, identify affected products and state thoughts on financial implications.
Cost	Thoughts on the cost of producing the product; identification of certain attributes of the idea that are particularly cost-sensitive.
IRR	Computation of an initial internal rate of return for the product.
Schedule	Information on the duration of the definition, planning, and execution phases. When each phase should start and a potential date for delivery of product to the customer should be indicated.
Risks	Shared thoughts on research and development, marketing, and manufacturing risks. The top two to three items should be identified.
Issues	The highlighting of items not fitting within previous categories.

Figure 2.2 illustrates a method of displaying the preceding information on a single page. This one-page document is called the new product definition sheet (NPDS). Note that space is provided for a graphical image of the idea and comments regarding all the remaining fields mentioned above. Some may envision it difficult to reduce a product idea down to a single page, but with persistence and some coaching, an individual or small group can develop the skills necessary to produce the NPDS.

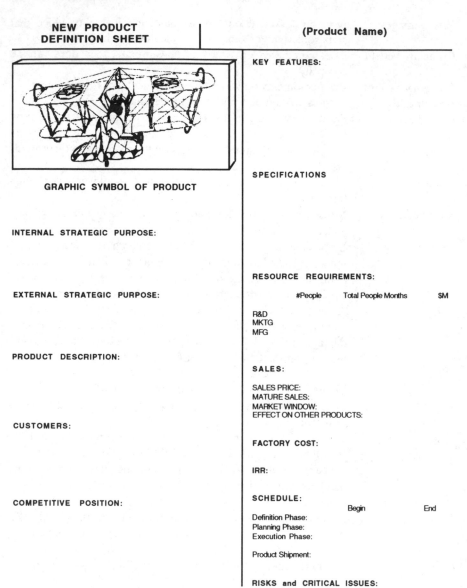

NEW PRODUCT DEFINITION SHEET	(Product Name)

GRAPHIC SYMBOL OF PRODUCT

KEY FEATURES:

SPECIFICATIONS

INTERNAL STRATEGIC PURPOSE:

RESOURCE REQUIREMENTS:

	#People	Total People Months	$M
EXTERNAL STRATEGIC PURPOSE:			
R&D			
MKTG			
MFG			

PRODUCT DESCRIPTION:

SALES:

SALES PRICE:
MATURE SALES:
MARKET WINDOW:
EFFECT ON OTHER PRODUCTS:

CUSTOMERS:

FACTORY COST:

IRR:

COMPETITIVE POSITION:

SCHEDULE:

	Begin	End
Definition Phase:		
Planning Phase:		
Execution Phase:		

Product Shipment:

RISKS and CRITICAL ISSUES:

Figure 2.2   Example new product definition sheet.

Also notice that each field represents an extensive unification of thought processes and requires the condensation and summarization of a substantial quantity of information. Once in existence, the NPDS becomes the vehicle to perform a first-tier comparative analysis with other new product ideas represented by their own unique NPDS. Think of the NPDS as the graphic symbol of the new product idea.

By representing each new product alternative with its own NPDS, the strategic business unit builds the capability to evaluate multiple alternatives simultaneously. This capability, contrary to some closely held notions, makes it possible to make quality decisions rapidly. The SBU is placed in the position where they can review, discuss, analyze, and improve upon multiple alternatives. They will often be working several options at once.

The SBU will find it valuable to categorize the NPDSs it receives. Commonly, a grouping by product line and by classification within a product line works to great advantage. For a given product line, the SBU groups me-too-with-a-twist, derivative, next-generation, and first-of-a-kind new product ideas. Once this is done, the SBU can begin to utilize the comparative process to float the best new product ideas in each category to the surface. In so doing, a rank order of new product ideas will be developed along with the reasons for idea placement on the list.

The SBU should develop answers to the following questions for its area of product and market focus:

How does the new product idea contribute to the company's vision?

How does the idea fit the "big picture?"

Does market information give a preliminary indication that the new product idea will be successful?

Does the NPDS symbolize a new product development undertaking that seems realistic?

What are the major opportunities presented by the idea?

What are the major problems associated with the idea?

At this point in time, how does the idea compare to other ideas?

The first three questions follow directly from the first three cells of the business column of the definition matrix. If corporate management has performed a satisfactory job in communicating the information content of the vision, big picture, and market information cells to strategic business units contained within the corporation, the information can now be put to use in making critical new product decisions. In effect, new product ideas will pass through a first-tier evaluation filter that is based on the first three cells of the definition matrix.

New product ideas that, in the determination of the strategic business unit, do not advance the corporate vision should be returned to the originators. This knowledge then makes it possible for the creators of the new product idea to take action. If these people are deeply committed to the idea, they may elect to champion its cause and attempt to help the organization make a conscious change in its direction. This is likely to be an arduous struggle, and one that arguably should meet with substantial resistance—a new product idea that moves a corporation in a direction different than its stated vision deserves to be thoroughly evaluated before making a change. Most probably, however, the individual or small group will be advised to develop the new product idea elsewhere. For the rare individual with the spirt to take this step, the process to this point will have at least partially prepared the entrepreneur for being alone in business without corporate shelter and support. Preparation for entrepreneurial business development is greatly assisted by having prepared the new product definition sheet. Those of you who have taken the step to develop a new product idea on your own will recognize that production of the NPDS forms an excellent starting point for the development of a complete plan for creating a business around the new idea. Of course, a new product idea consistent with the corporate vision moves on to be evaluated against corporate objectives.

Attention shifts to developing an understanding of how the new product idea contributes to the guiding objectives of the corporation. It's reasonably easy to decide what to do with an idea that has a negative impact on profitability, growth, customer satisfaction, people within the organization, the surrounding community, or the environment. Furthermore, if higher-level objectives exit, the new product idea must not negatively impact these admirable intentions. A new product idea found to support global corporate objectives can then be viewed to determine its impact on time-related objectives. The strategic business unit determines the effect development of the new product would have on the two- to five-year objectives. It's important to see if a new product idea fits into what the organization is attempting to achieve. If the complementary nature of the idea cannot be established, it's likely the SBU will return the new product definition sheet to the idea creators for continued development. An idea that appears to fully support and enhance time-related objectives is ready to be examined from the "big picture" perspective.

It's possible, within a relatively large corporation, for a new product idea, congruent with the corporation's vision, to surface within a business unit attending to differing areas of interest. Knowing the "big picture" makes it possible for an SBU to transfer the new product idea to the product-line managers who are likely to have the most interest in the product.

Assuming the idea is now resident within the best-fit strategic business unit, time and energy can be applied to better understand the idea's impact on existing products and services, customers, channels of distribution, and the competition. At this point, the comparative process of evaluating new product alternatives reveals more of its advantages. Certain ideas start to look significantly better than others. At times, segments of one new product idea can be applied to improve upon another. The members of the SBU begin to compare ideas against one another to improve understanding and begin forming educated opinions of relative strengths. New product ideas that are seen to increment the product portfolio significantly, meet customer requirements, work within the existing distribution system, and make a competitor's ability to conduct business more difficult are more likely to gain the acceptance of the SBU than those new product ideas that are lacking in certain "big picture" attributes. As the SBU begins to identify areas of weakness within particular new product ideas, it is vital to get this information back to the creators of the idea. This candid assessment can sometimes give the creators the stimulus needed to turn a good new product idea into something that radiates greatness! It's likely they might think of ways to circumvent the identified problems by making creative adjustment to product and process. As negative attributes are removed, SBU support for the idea is likely to expand. Sometimes the added effort that went into idea improvement elevates its importance within the SBU. When inspiration does not result, it's probably best to just leave the idea at rest for a period of time. New product ideas that pass through this filter can then move on for evaluation from the final, first-tier perspective.

Information on the new product definition sheet can now be evaluated against information contained within the product, customer, competitor, and market databases. It now is possible to become more objective. Subjective impressions start to give way to the production of relative facts and figures. The SBU begins to do a less than, greater than, and about equal to analysis. Effort can be extended to document if the product represented by the NPDS will sell less, more, or about the same as existing products. Growth rates can be examined in the same way and compared with the growth rates of the market and the impact on competition to develop a feel for gain, loss, or parity in market share. Also, the SBU can look to see if the product solves documented problems in existing business operations. Likewise, preliminary assessments of customer response can be made by the SBU. The relativistic data can be put to leveraged utilization to arrive at the identification of the new product development ideas that should move on to the analysis cell of the definition matrix.

Concurrent with the evaluation of the NPDS from the definition matrix perspective is the background development of answers to the questions that pertain to realism, opportunities, problems, and relative placement. SBU members become the integration point for what is commonly too much information in one area and too little in another. The SBU must be skilled at data reduction in cases where the data reserve is excessive and be proficient at working in other areas in which moving forward with limited supporting evidence is essential. To work in an environment of this type requires that SBU members typically be older and more experienced. They will have worked within the organization and the industry for a period of years to refine their skills. These individuals must be recognized as having savvy or being "street smart" in their particular areas of expertise.

An excellent composition for an SBU is usually one with key individuals from research and development, marketing, and manufacturing. Ideally, each participant not only has a good understanding of the entire operation of the organization and knows how other functions contribute to the success of the enterprise, but also brings to the team a particular area of expertise that may be held by only a few other people within the industry. The research and development person may understand how marketing and manufacturing work from a high-level perspective and also possess unique expertise in devise technology, design systems, and new product development process management. The marketing participant's special strengths may focus on knowledge of key customers, product applications, and worldwide distribution techniques. The manufacturing member's unique skills may focus on design for manufacturing systems, subcontractor relations, and labor relations. Integration of the global skills and the areas of strength makes for a powerful team structure. The depth and breadth of business understanding resident within the team are exactly what is needed to make the critical judgments about each NPDS and produce the relative ranking of new product alternatives.

As the SBU becomes comfortable with each NPDS under consideration, a process of integration begins. Alternatives that seem unrealistic from a number of perspectives begin their descent on the list. Those that appear to offer ample opportunity with a problem set that seems workable begin the ascent. Those in the middle tend to stay in the middle. Conflicts will develop about various issues. The size of the SBU and the close working relationships make conflict resolution a speedy process. The group works to reach a consensus. Where decisions are viewed to be time-critical and the SBU is having difficulty developing a consensus, a previously appointed SBU member is called on to make the decision that maintains forward progress. The high-stakes decisions made by the SBU link the ideas under consideration with each other and with new products currently under development. Integration of decisions gives the SBU a degree of control over the process. Things begin to make sense. Each alternative becomes better understood with respect to the others. Views on how to set strategy, tactics, and logistics for the development of favored new product ideas begin to take shape. The SBU develops a strong preference for specific new product ideas.

If, based on SBU reasoning and preliminary supporting data, the idea fits the basic needs of the organization, or it looks like changing the organization to fit the new product's developmental needs is not out of the question; if the market is there and is big enough; if the organization has the skill, or can acquire the skill, to develop, produce, and distribute the product; if the idea looks reasonable from a financial perspective; then the SBU must take action and assign a small staff to develop each of the selected new product ideas further. However, before moving to detailed alternative analysis, the SBU must fully communicate its findings to the creators of the ideas and to the management team of the organization to ensure that all necessary process participants are aware of, and mostly in support of, the decisions and recommended actions.

Figure 2.3 presents a checklist of items to consider during the first tier of filtering new product ideas while in the alternative synthesis stage.

Individuals who generate new product ideas and document them on a new product definition sheet are encouraged to use the checklist to critique their own work. Strategic business units are encouraged to use the list as a guide. In addition, the SBU may want to utilize sections of the checklist as a communication vehicle to the originators of the new product idea and to facilitate discussion with the management of the organization.

2.1.5   Business Alternative Analysis

Of the many possible new product ideas that could be entertained by the organization, only a few select ones have survived to this point.

The purpose of alternative analysis is to investigate selected synthesis ideas thoroughly from a complete business perspective before the commitment of staff and large sums of money. The new product idea must be evaluated from the standpoint of the organization's ability to produce the product and from the level of financial contribution the product is expected to yield. It is the second step in the process of new product idea selection and refinement where people will be tasked with doing a detailed investigation about the idea's merits. At this point, the strategic business

_____	Good alignment with corporate vision.
_____	Supports overall corporate objectives.
_____	Supports time related corporate objectives.
_____	Fits SBU area of expertise.
_____	Positive impact on product and service offering.
_____	Good alignment with customer wants/needs.
_____	Works within existing distribution system.
_____	Presents a significant challenge to the competition.
_____	Position on Product Classification Matrix is understood.
_____	Close alignment with organization's areas of expertise.
_____	Certain aspects of the idea are elegant and simple.
_____	Workable from a research and development perspective.
_____	Workable from a marketing perspective.
_____	Workable from a manufacturing perspective.
_____	Aspects of new product development strategy, tactics, and logistics are visible.
_____	It is the right time.
_____	The market appears to be there.
_____	Looks like the organization can respond early in the market window.
_____	Has the potential to generate an impressive ROI.
_____	Has potential to spin off in other directions.
_____	Resides at or near the top of all ideas presently under consideration.
_____	All members of the SBU are in support of the idea.
_____	Some risk reduction strategies can be identified.
_____	Suggested before in a slightly different format.
_____	People with a history of producing winning products want to work on the development team.

Figure 2.3 Business alternative synthesis checklist.

unit is likely to appoint key individuals to analyze the remaining alternatives fully and investigate key aspects of each idea that appear to present the greatest new product development challenges.

Research and development will gather in-depth information on technical approaches, technical risk, staff estimates, time-to-market estimates, and risk factors. In high-risk areas, a researcher might be assigned to explore a particular area of elevated uncertainty. Marketing will work with the research and development team to generate preliminary performance, feature, and price attributes of the new product along with early estimates of introduction timing sensitivity and market window parameters. Once early conjectures begin to stabilize, marketing can then look into the sales potential of the new product idea while R&D refines its estimates in view of the new information that has been produced. Manufacturing will usually wait on the sidelines for a brief period of time until the product concept reaches an early set point. When it does, the manufacturing team can make its input about design for manufacturability, concept modifications, and estimates of manufacturing cost. Data from this team activity are input into financial models to arrive at return-on-investment (ROI) numbers for the new product idea.

Staff who perform this type of work usually have a vested interest in the continued success of the idea. More often than not, the small team, ordinarily three to five people, will be extremely skilled specialists from R&D, marketing, and manufacturing who have experience in doing the preliminary development and analysis of new product ideas. Their enthusiasm, or lack thereof, as the analysis task proceeds is an excellent test for the quality of the new product idea. This team of internal experts runs the idea through its second-tier review.

Each idea selected for analysis is assigned to a dedicated team. Team members will likely do this work while performing other job responsibilities. If structured correctly, the analysis team will be working with a number of selected ideas to continue the process of comparing ideas with one another. Progress on the analysis is fed back to the strategic business unit on an informal basis. The SBU will know from interaction with members of the analysis team the direction the analysis is taking. Enthusiasm and support for the best ideas tend to build by discussion and interaction well ahead of any written documents. As this task nears completion, the analysis team will prepare an updated edition of the new product definition sheet. This final NPDS accompanies the idea as it moves toward the definition phase and the product column of the definition matrix. An example of the updated NPDS is contained in the Appendix ("Example New Product Definition Sheet [NPDS]"). This simple document becomes the working symbol for the new product idea. It becomes the communication and descriptive vehicle to initiate the development of the new product as the team is assembled and assigned to take the effort forward.

The strategic business unit collects the feedback from all of the analysis teams and integrates this information into the new product strategy for the business on a frequent, periodic basis. As new information becomes available, the strategies are revised. This process is always in motion. SBUs receive information about each of the potential new product alternatives on a daily basis but rely on the NPDS as the stake holding the idea to earth. For each alternative, data about customers, competition, product performance and features, product price, resource requirements, time to market, market window, schedules, investment cash flow, and returns are generated. The SBU must now begin the process of blending the selected new product development ideas into the overall operational capabilities of the organization. This is no easy task! The SBU will have to select alternatives that allow the organization to function within its means while producing a very respectable financial return.

Architectural aspects of the product, which are going to determine the total work effort required to get the product to market, are now beginning to influence the effort. The most important early test a new product idea receives is the preliminary determination of its time to market and its market window for the architecture under consideration. The strategic business unit must look at these figures and determine, if the project is to proceed, how the organization can go about responding in this time frame and if it has the resources and funds to do so. Viewed from this perspective, the ideas that look achievable start to be distinguished from those that appear to be difficult. The SBU views the timing, resource, and cost information on the new product definition sheet in a collective fashion and asks critical questions. Do the time-to-market and market window numbers seem reasonable? Does it look like the estimation of resource requirements was done on an informed, cross-functional, basis? Are development costs accurately stated? The SBU and analysis teams are certain to interact on these issues. Developing a level of comfort with the process and the figures requires close contact between participants as understanding grows.

This point of view ensures that the strategic business unit rejects new product ideas that won't make it to the market on time. In effect, early in the process attempts are made to align the timing requirements of new product development with the timing signals being received from the market. This point of reference helps the organization from starting a project too late, or it can force an architectural revision in the new product idea that could speed the product development process. Looking for architectural ways to speed the product to market or putting late entry ideas to rest at this point in time are wise actions.

Those alternatives that look feasible, if appropriately staffed and funded, are then integrated into the overall flow of new products from the organization. In this case, the strategic business unit examines organizational capacity to determine if a new product development effort can be executed within the time frame set by the market. This analysis requires the SBU to view the development of each new product based on the phased availability of existing resources and the ability of the organization to add new resources. The flow of resources to and from new product development efforts must be modeled and analyzed at periodic intervals over the year. In effect, the SBU must overlay and phase staffing profiles from each new product development effort to ensure that resource peaks do not exceed capacity and that timing for each product can be accomplished within the appropriate market window.

The strategic business unit and analysis teams that do this job realize the exercise forces the production of difficult decisions. When confronted with the possibility of not pursuing certain efforts because of resource limitations, the first temptation is to run all new product development efforts at reduced levels of staffing commitment. To do this, however, means that full staffing is never achieved on any one of the projects and all will have delayed market entry. This thinking style will lead to eventual business misfortune! What's needed is the forced selection of the best new product ideas that can be fully staffed and presented to the market on time. Making the difficult decisions about which new product efforts to staff fully and get to market becomes a primary function of the SBU and the organization's top management team.

To help clarify concepts about activities taking place within this cell of the definition matrix, I'll draw from a situation I was in a few years ago.

For discussion, let's suppose a strategic business unit assigned independent analysis teams to study three new product ideas. Next-generation, derivative, and first-of-a-kind product ideas were proposed and have passed the first-tier screening. The team selected to analyze the next-generation proposal consisted of three highly rated design engineers with many years of systems and technology experience, a marketing engineer who had developed the market when the product classification was first-of-a-kind a few years prior, and a manufacturing engineer currently supporting the commercial manufacturing assembly process. The new product definition sheet of the derivative product reflected the overall simplicity of the endeavor. An experienced manufacturing engineer was united with the person who generated the new product idea. A marketing engineer with only a few years of experience was added to round out this team. The first-of-a-kind team was assembled from research and development, marketing, and manufacturing people with a track record of success on a previous first-of-a-kind project. In addition, two research engineers who had developed the initial product concepts were added to this analysis team. The organization was asking a group of highly talented people to help it decide where to focus its new product development resources.

As might be expected, the derivative team was the first to complete their analysis. They found that the investment to bring this idea to market was small in comparison to the returns it would generate. At the heart of the idea was a 25-percent reduction in manufacturing costs. This gave the organization the option to retain existing prices and increase its profitability, or be more aggressive on pricing and increase market share in the present market while the new derivative product opened up new areas of opportunity. This was just too good an opportunity to pass up. Support for this idea began to grow. Developing the product and getting it to market swiftly was of paramount importance—a late entry provides minimal rewards. Staffing and expense requirements to undertake this effort were certainly within the organization's capabilities.

As is typical of next-generation product ideas, design and process improvements began to appear harder to realize than initially thought by those who generated the proposal. While the investigating staff size was small, the experience and capabilities of the team assigned to this analysis and study effort were substantial. Research and development people were having trouble making any major technological advances. Consequently, the marketing representative could not get excited about the small gains in performance that the new product would offer over the existing product. Manufacturing saw more opportunity in reducing the cost of current products than it did in the untried processes being proposed for this new product idea. Upon detailed study, even the R&D members of the analysis team began to lose their enthusiasm. The entire team began to form the opinion that this idea would be more work than it was worth, but were reserved about directly saying so. The team's return on investment analysis showed that the internal rate of return (IRR) for this proposal was very low.

Acceptance of first-of-a-kind ideas tends to run on a hot-to-cold cycle. Analysis of this proposal started at the cold point of the cycle and stayed there for a period of time. The idea was loaded with problems and uncertainty. The research people had demonstrated they could produce a crude, but functioning, laboratory sample. They also demonstrated that the manufacturing processes required to mass-produce the design were not yet available. Marketing saw the laboratory sample and began to fantasize about having 100-percent market share of what was thought to be a large market for a long period of time. The link between this product and dominance in a new market was very visible. The manufacturing representative solicited outside expertise and added two additional members to the team. The R&D people and the manufacturing people began to uncover methods that could make the design producible. They agreed that the product design and process design would have to be simultaneous to meet time-to-market constraints. Also, they were able to highlight the areas where risks seemed the greatest. All members of the team were convinced that if they could get a manufacturable design to market, they would have created one of the highest IRR projects in the organization's history! Timing to bring the product to market, while important, was viewed to flexible by at least one or two years.

As the strategic business unit became familiar with the above alternatives, certain parameters that would influence their eventual decision started to surface. It became obvious to the members of the SBU that sufficient resources were not available to do all three projects simultaneously—to do the next-generation effort and the first-of-a-kind effort concurrently would not be possible. Two of the three product efforts were headed in the direction of new markets. As such, there was an apparent lack of new products to service existing markets. Also, there were a number of me-too-with-a-twist ideas available for current products that, up to now, had not been given serious consideration.

Utilizing the information presented, what do you think the SBU should have done?

This particular group of people at this point in time produced the following decisions:

1. Move the derivative product development effort to full development status. Make it a formal project and begin its definition phase. As the definition phase progresses, be prepared to support the project with resources and money to keep it on time.
2. Terminate all work on the next-generation effort. Watch the development of new ideas, but to maintain and possibly grow in current markets served, switch to a me-too-with-a-twist incremental improvement strategy. This "smaller step" strategy is expected to give the product developers the opportunity to improve the product offering in small amounts but upon frequent occasion, thereby lowering risk and reducing resource requirements. Prepare to apply up to 25 percent of engineering staff to incremental improvement projects once the new product definition sheets have been prepared.
3. Assign remaining available resources to the first-of-a-kind effort. Start the definition phase. Make certain this team knows they are third on the priority list for resource and financial support, but also let them know their efforts will have a major contribution to the vitality of the organization in a few years.

The strategic business unit, having completed the analysis of alternatives and making up its collective mind about what should be done, is now in position to collect the information it has available and make a full disclosure and recommendation to the organization's top management team.

To finish exploring the contents of this definition matrix cell, please turn to the MONI Display Division Business Definition document in the Appendix (under "Division Business Definition Example").

As a divisional work entity, the SBUs within the display division reduced the number of new product ideas to receive staffing commitment from a relatively large number down to six. The six new products under active consideration are referred to as Better Lookman, Personal Display, 3/4 Cost, Revolution, Rainbow, and U-Tube in the document. A brief description of each product is contained in the section titled "Necessary Products and Services." Under "Development and Introduction Plan" section, the business plan documents when each of the new products under development is expected to be available for volume distribution. Each product mentioned in the document has a corresponding new product definition sheet that becomes an integral element of the plan. The NPDS for the Revolution product is supplied for your reference ("Example New Product Definition Sheet [NPDS]"). The display division also prepared a resource profile for each of the six product development efforts and merged them over a period of time on a spreadsheet shown in the "Development Staff and Costs" section of the document. As you review the staffing profile of each product, note the controlled growth in the summation fields of research and development, marketing, and manufacturing. If you like, a graphical construction of staffing profiles may be of value to help see how people are expected to be moving between projects. You'll also note, within the last six months of this two-year window, a number of resources become available under the field labeled "Other." This indicates that the display division's product development portfolio loses clarity the farther the plan looks into the future. The last item in the document associated with this cell is the "Potential Problem Analysis." Here the SBUs and analysts point out concerns they have about each of the recommended development efforts.

It's important to remember that the required work to complete this cell was performed by the strategic business units and by highly skilled lead people in the organization. Prototypes of the product have not been built. Only simple simulations and visualizations of the product are in existence—to this point, it has largely been a mind game. The people involved have done their best to predict the future. By now, the initial architecture of the product will have been established, performance goals will have been stated, features will have been crafted, and a wealth of numerical, textual, and graphical information will have been assembled. The stage for what is about to be revealed over time has been set. In certain areas, the people who worked on the analysis will be right, in others . . .

None of the work traditionally performed by product development teams has as yet been initiated. Commiting the organization to enter the definition phase on selected product alternatives occurs in the next cell of the definition matrix.

2.1.6 Business Output

The business definition process is nearing completion. As you have probably noticed, contribution to the creation and content of the business definition requires participation from most individuals in top management positions and from key research and development, marketing, and manufacturing professionals. This work takes place in the upper segments of the organization's hierarchy but embraces all horizontal aspects of the business. Corporate management should be comfortable providing the vision and big picture for the overall corporation. This group can set the wheels in motion to ensure that systems are in place to collect vital market information. Of course, it can assist in the integration of plans between groups and product divisions. Lastly, corporate management can monitor the performance of the new product development effort against communicated intentions and use the information it collects to keep the efforts on track or make adjustments when necessary. In a large, multidivisional firm, corporate management is not in a good position to work directly with alternative synthesis and alternative analysis cells of the definition matrix business column. The strategic business unit is best equipped to work these cells.

At the division level, assuming SBUs exist within these entities, a complete business definition is possible. A division can supply information on the entire business column of the definition matrix. All that is necessary is that the division management team provide input on the vision and big picture to the SBUs that operate within the division. The SBUs then complete the remaining four sections of the business column. Once the output from each of the SBUs has been accepted, the division management team can fully integrate all of the inputs and produce a business definition for the entire division.

Once again, please refer to the MONI Corporation Business Definition document and the MONI Display Division Business Definition in the Appendix (under "Corporate Business Definition Example" and "Division Business Definition Example," respectively). These documents are example deliverables from the business output cell of the definition matrix.

In the MONI Corporation Business Definition, the "Vision," "Five-Year Objectives," "Customers and Channels of Distribution," and "Competition" sections were produced by the top management team of the organization. As you read the information they provide, ask the question, Do I understand MONI's vision and do I comprehend the "big picture?" In the next section, "Necessary Products and Services," MONI top management tells its divisions and SBUs where the corporation wants to focus its product development efforts and then follows with a "Market Information Summary" that links product lines to markets and expected financial performance. The "Product-Line Guidance" section provides control information to MONI's divisions and strategic business units. Top management provides this feedback after carefully reviewing operational, technological, customer, and market information. As you read the comments made about each product classification, note that MONI top management is not afraid to take a position on key product-related issues. As might be expected, information from divisions and SBUs was used to help make the decisions that led to the production of these guidelines. The final section of the document highlights problem areas that are of concern to top management. This document, once supplied to MONI's operating divisions, leads to the production of their business documents.

The display division extracts information from the corporate document to produce the MONI Display Division Business Definition. Note the "Vision," "Five-Year Objectives," and "Customers and Channels of Distribution" flow from the corporate document to the divisional document with little modification. A linking of this type is highly desirable—the upper operating levels of the company are firmly united. Not surprisingly, the division's competitive view becomes much more specific than the corporate perspective. Each division in MONI will address certain competitive situations that directly affect their business. If each business unit of MONI does its job well, the corporation will present ample challenge to its competitors. Each division then identifies the new products it presently has under development and lists those it has selected to move to definition phase status once resources are available. As previously discussed, the work performed in the synthesis and analysis cells led to the production of these new product decisions. Each division then shows when products should be available in the "Development and Introduction Plan" section. If further information about any of the selected products is needed, the document contains a new product definition sheet for each product (only one follows the document in the Appendix). Resource and cost information follows to illustrate how people and money will be used to get the new products to market. Division management then has to merge and fully integrate the response of the three SBUs that operate within the entity. The "Recommendations" section documents the decisions made by division management as they integrated each SBU's plan into the overall operations plan. Finally, now that each SBU knows the rules, it becomes possible to produce a brief, high-level tactical plan for the division.

The corporation is now in the position to float bottom information up and sink top information down. As the data move vertically, dispersion effects will be at work to move it horizontally as well. The net result is that the entire organization now knows how to initiate the new product development process effectively.

The business definition process that has been described works best when

1. Corporate, division, and strategic business unit managers make the development of new products a very high priority and place strong emphasis on business definition within the corporation.
2. Strategic business units are managed by a small staff who are talented and who have demonstrated their strategic synthesis skills.
3. Strategic business units have a common structure with which to view the market.
4. Strategic business units are actively involved with customers to understand their *business* problems.
5. Strategic business units have a firm grasp on technology and its limitations and applications.
6. The corporation has a process established to create, review, and update business definitions.

The business definition process works its way vertically and horizontally throughout the entire organization. Business definitions are produced to simplify and improve the process of creating product definitions. New product development teams now have a clarity of focus that did not exist prior to performing this multi-level, cross-functional activity. This focus helps to limit development complexity, reduce the occurrence of negative conformity, shine light on some invisible items, reveal some of the true development costs, make visible the preliminary resource requirements, identify time-to-market impediments, and stimulate the "can do" spirit of the development team. Nothing is hurt by the establishment of a business definition, whereas just about everybody (idea people, new product development teams, business teams, division management teams, and corporate management teams) within the corporation has something to gain.

To initiate work on the product column of the definition matrix, form a core team and hand them the new product definition sheet . . .

3

PRODUCT DEFINITION

In this chapter, you will learn about

The new product development process (NPDP) matrix.

The creation of internal and external product visions.

The generation of a "big picture" for the product.

The product development application of market information.

The synthesis of product alternatives.

The analysis of product alternatives.

The output of the product definition process.

Problems that can be solved or minimized by the application of good definition techniques.

Relationships that can be used to estimate hardware time to market.

Relationships that can be used to estimate software time to market.

The contributions of the definition phase to the overall process.

Imagination is more important than information.

Albert Einstein

3.1 PRODUCT DEFINITION SYNTHESIS

Output from the business definition process is transferred to new product development teams. A definition team has available to it a corporate definition, a division definition, and the preliminary new product definition sheet for its project. Of the three, the primary initiator of the new product's development is the NPDS. The other documents provide supporting information that is necessary for team success. With the preliminary NPDS complete and resources committed to begin the definition phase, the SBU and top management ask the new team to develop the definition of the product thoroughly. In effect, the NPDS and a small staff are all that's required to get a new product development effort started. Hopefully, some of the individuals that worked on the NPDS in the business column of the definition matrix will be available to work on the effort now that it has transferred to the product column. This group, which consists of individuals from many of the organization's functions, is now in the position to assemble the needed information to produce a complete product definition that satisfies the horizontal and vertical information requirements of the organization. In all but the smallest of companies, it is likely that a number of new product development projects will be occurring simultaneously. The timing between projects will be managed to keep the projects working within organizational resource constraints.

Overall, the product definition can be segmented into the six major topics of product vision, big picture, market information, alternative synthesis, alternative analysis, and output. These are the cells arranged along the product column of the definition matrix in Figure 3.1.

In the previous chapter, focus was on the business column. This chapter will detail the contents of each definition matrix cell in the product column. This journey will present the most demanding aspect of new product development: defining the exact nature of the work the development team will perform. When finished with the product column of the definition matrix, the product development team will have a clear, unified view of the product development effort that lies ahead. The development team will know it is working on the "right thing!"

3.1.1   Product Vision

The product vision is intended to acquaint product development team members with the internal business purpose and the external purpose of the product. First, focus is directed toward the internal purpose. A series of example internal product vision statements follows:

> A next-generation product development effort to simultaneously improve the quality, cost, and time-to-market image of our company.

In this product vision, the company states that it wants to become a high-quality producer, low-cost manufacturer, and rapid developer of new products. Doesn't everyone? How many organizations have the foresight to use the new product development effort to achieve this strategic objective? Achievement of these objectives is likely to contribute to the long-term business success of this company. It is interesting that a new product development effort is being used to simultaneously launch initiatives that will affect the entire organization. People working on this effort know, without doubt, what is expected of them. Hewlett-Packard has followed this strategy remarkably well in its printer product offerings.

> A derivative product effort to respond to the feedback that our customers have supplied about our existing product.

To respond to customer feedback, new product strategists chose to utilize a derivative product effort to meet this challenge. With the derivative product architecture, it is likely that they will be able to modify the product's design easily over time to adapt it exactly to customer requirements. They are also establishing a foundation for the rapid development of future new products that will be based on the derivative architecture. Future business success will be based on the ability to convert cus-

	BUSINESS	PRODUCT
VISION	VISION OF THE BUSINESS CORPORATE OBJECTIVES 3 TO 5 YEAR OBJECTIVES	INTERNAL PRODUCT PURPOSE EXTERNAL PRODUCT PURPOSE
BIG PICTURE	NECESSARY PRODUCTS NECESSARY SERVICES TARGET CUSTOMERS CHANNELS OF DISTRIBUTION COMPETITIVE PRESSURES	PRODUCT DESCRIPTION TARGET CUSTOMERS COMPETITIVE POSITION
MARKET INFORMATION	CUSTOMER DATABASE PRODUCT DATABASE COMPETITOR DATABASE MARKET DATABASE DATA REDUCTION	NECESSARY PERFORMANCE NECESSARY FEATURES NECESSARY COST NECESSARY PRICE NECESSARY TIMING
ALTERNATIVE SYNTHESIS	DESCRIPTIONS & SPECIFICATIONS TECHNICAL APPROACH TECHNICAL RISK INVESTMENT TIME TO MARKET	INTERFACE DESCRIPTION SPECIFICATIONS TECHNICAL RISK INVESTMENT AND TIMING FIT WITH PRESENT SKILLS
ALTERNATIVE ANALYSIS	IMPACT ON THE MARKET RETURNS FROM THE MARKET FINANCIAL ANALYSIS POTENTIAL PROBLEMS	IMPACT ON THE MARKET RETURNS GENERATED DETAILED FINANCIAL ANALYSIS POTENTIAL PROBLEMS
OUTPUT	CORPORATE BUSINESS DEFINITION DIVISION BUSINESS DEFINITION PRELIMINARY PRODUCT DEFINITIONS STRATEGIC RECOMMENDATIONS	PRODUCT DEFINITION NPDP MATRIX ENTRIES COMPLETE RECORDED ESTIMATES

The left margin reads vertically: **D E F I N I T I O N   M A T R I X**

Figure 3.1 Definition matrix.

tomer input into product reality rapidly. Porsche follows this strategy with its 911 Series of automobiles.

A derivative effort to create the formation of a new market niche.

An existing product will be modified to establish a new market. Since the development costs and time to market to enter the market with a modified product are less than a full-scale development effort, this strategy represents a lower-risk approach to market development. The product can be thought of as a probe—future action will be based on what is learned from the market. Ford used a derivative of a pickup truck, the Bronco II, to test if the urban 4 × 4 market was a reality.

A me-too-with-a-twist effort to grow within a rapidly expanding new market.

A new high-growth market is emerging. Opportunity is knocking at the door. Strategists have decided to participate in the market and want to develop a product that will be competitive. Since the market is in the formative process, a competitive product that is rapidly developed is a must for this effort to succeed. Where did all the IBM clone developers come from?

A next-generation product development effort to reduce our dependence on a large number of material suppliers and establish a new manufacturing capability.

Reducing part count in the design of products with a corresponding decrease in parts suppliers has proved its merits as a competitive product development strategy. The present-generation product will be engineered to reduce part count significantly and thereby reduce manufacturing complexity. Moreover, those fewer parts will be produced by still fewer vendors. Just-in-time (JIT) manufacturing processes will be established to reduce manufacturing cycle times and virtually eliminate inventory. Suppliers will be tightly integrated into the manufacturing process. Xerox used this strategy to reestablish itself in the copier market.

A me-too-with-a-twist product development effort to utilize global technology as a strategic business force multiplier.

Emerging global technologies have the capability to transform the product development, manufacturing, and distribution processes. This statement tells the development team that it is okay to look for the best the world has to offer and then go ahead and use it! The new product will be designed so that it makes optimum use of the new developments in all facets of its business life. General Motors is following this strategy with its Saturn automobile project.

A next-generation new product effort to develop our skills at doing concurrent new product development while making our organizational structure less complex and more responsive.

These are the characteristics of the lean and mean organization except that a new product development process is being created from the bottom up. This organization realized that a deep and complex hierarchical structure is not conducive to the concurrent development of a new product. A new structure, modeled with the use of matrices and networks, will be built around the job that is to be done, rather than fitting the job to the organization. Ford's development of the Taurus followed a strategy of this type. Benetton has gained world dominance in their chosen markets with a flat organization.

A me-too-with-a-twist effort to develop new design for manufacturing (DFM) fundamentals.

Leadership manufacturing skills have shown their competitive merits. Hitachi and General Electric had early success with DFM and promoted the process they had developed. In the mid-1980s, the work of Geoffrey Boothroyd and Peter Dewhurst stimulated the DFM movement. Adoption of this strategy continues to gain momentum. Combining world-class manufacturing with effective product design results in an unbeatable strategy. The above statement tells the members of the new product development team that it is no longer acceptable for the research and development players to throw the design over the wall to manufacturing. It's a major first step in getting cross-functional, horizontal communication initiated. Success has been demonstrated by IBM (printers), Texas Instruments (night-vision products), NCR (cash registers), and Ford (the Taurus line of cars).

The list below highlights the concepts behind an internal product vision statement. Examination of the examples reveals that the product vision has properties somewhat like the business vision. The seven criteria used to evaluate an internal product definition follow:

1. It classifies the type of development effort being undertaken.
2. It does not speak of product performance, features, or price.
3. It provides a strategic reason for doing the project beyond the obvious one of making money.
4. It tends to challenge the development team.
5. It's clear, understandable, and nontechnical.
6. It places strategic constraints on the product development effort.
7. It looks inward.

The statement that follows is extracted from Silicon Graphics' 1990 annual report. Let's use the above criterion to evaluate the statement's effectiveness as an internal product vision.

This next generation . . . supports concurrent engineering—where 3D data is accessible and concurrently used by designers, engineers, materials analysts, and marketing and accounting specialists to get higher quality products to market at an ever faster pace.

First, the statement starts out with the words *next generation*. This leaves no doubt about the classification of the development effort being undertaken. Second, no mention is made of product performance, features, or price. Third, Silicon Graphics states their strategic reason for doing the product—they want to get higher-quality products to market at an ever-faster pace. Fourth, the challenge to the development team is stated. Making three-dimensional data available to this diverse a group is no easy task. Fifth, the statement is easy to understand—even in a technology-intensive company such as Silicon Graphics, it is possible to describe a product development effort simply. If you are working on this project, it's very clear why it's being done. Sixth, strategic constraints have been stated. The product is intended to enhance the internal concurrent engineering process. Seventh, this vision looks inward. This product development effort is intended to improve the new product development process at Silicon Graphics. In all, this statement is a high-quality internal product vision statement.

In contrast, consider a typical internal vision that forms within many organizations:

A product that generates a lot of sales so that the organization can meet its profit and growth targets.

As a member of the new product development team, you have just been told what you already know. The statement does not classify the type of development effort being undertaken; it does not provide a strategic reason for producing the product; it does not challenge the development team; it is not clear and understandable; and it does not place constraints on the product development effort. Statements of this type are poor from an internal vision perspective. Remember, if the product under development were not a money-maker, you would not be doing it!

Before leaving this topic, please turn to the new product definition sheet of the MONI 2001 flat color display in the Appendix ("Example New Product Definition Sheet [NPDS]"). MONI's internal product vision for it is: "A next-generation development effort to keep MONI technologically ahead of the competition and retain market dominance. A display technology for all future home and computer products." This statement makes perfectly clear why MONI wants this product! MONI is a large corporation that cannot afford to yield technological superiority in the marketplace. Strategically, MONI has decided that it is not going to buy its next-generation displays; it will develop them internally. As you test the MONI statement against the seven product definition statement criteria, it becomes clear that all are satisfied. Additionally, the creators of this statement wanted to impress upon the development team the leverage their work would have throughout MONI.

It is instructive to gather a few internal product vision statements used in your organization and objectively review their message content. If written statements do not exist, try to reconstruct verbal samples.

How many of the characteristics are satisfied? A fragmented or nonexistent internal product vision indicates that your organization needs to make improvements in this area.

To be effective, a new product development team needs to know why its work is strategically important to its company. The internal product vision gives it this information.

Furthermore, the team needs to know why its work will be important to the customer. Correctly structured, the external vision of the product leads to, or sometimes is, the position statement for the product. Robert Heller identified the "seven satisfiers" in his book *The Decision Makers*. When a product fills an unfilled need, removes a disadvantage in an established product, fills a gap in an otherwise well-served market, extends or provides new formats within a given line, makes a technological breakthrough, transfers success in one market to that of another, or provides an economical way of satisfying needs now being expensively met, then it is doing something useful for the customer. Such a message of position and service must be stated—as viewed from the customer's perspective—in the external vision statement. A few examples to illustrate the concept follow:

This product

- Saves the customer money:

Now you can predict system performance before costly implementation.
—Signal Technology Incorporated

Why did we introduce a workstation for $4995? Because we could.
—Sun

- Gives the customer a new capability:

You can work across two full pages with the photographic quality of 24 bit color.
—Radius

Imagine a single network that could allow half the world's population to talk to one another—all at the same time.
—BNR

To grasp its true power, get your hands on one at your nearest HP retailer.
—Hewlett-Packard

- Makes it easier for the customer to do something:

Another improvement designed to make Mac enthusiasts even more enthusiastic.
—Microsoft

Announcing a cure for the terminal blues.
—SAS Institute Inc.

Designing on any other workstation will seem downright primitive.

—IBM

* Improves the customer's skills at something:

Because a document should reflect your aspirations not your size.

—GCC Technologies

To go anywhere in engineering you have to know what's going on around you.

—Dialog Information Services

* Teaches the customer something.
* Corrects problems.
* Extends the capabilities of something.

The preceding examples reflect the enthusiasm of the advertising departments that produced them, but still convey the product's purpose to the user. As part of the product definition, they probably started out life in bland terms. Nonetheless, a statement about the product's external purpose tells the development team how the customer will benefit from its work. If many benefits are possible, it is best to keep the statement focused on a few key items. To produce a statement of your own, start with the bulleted words. To them, add the necessary structure to create the external product vision. Over time, an accurate but simply stated external vision will transform itself into a quality customer statement.

Returning to the MONI new product definition sheet in the Appendix ("Example New Product Definition Sheet [NPDS]"), the external purpose of the product is given as

A new video presentation screen that provides unsurpassed information display capability.

This statement conveys the message to the development team that they are providing end users with display capability that will improve their customer's viewing of information. The statement should motivate those who have a sincere interest in doing this type of work. It provides the developers with yet another clear piece of strategic information to keep them focused on the main job to be done.

3.1.2 Product Big Picture

In this cell of the definition matrix, the overview information about the product receives more attention to detail. The development team refines the information associated with the product description, customers, and competitors provided on the preliminary new product definition sheet. Cross-functional team members will now begin to interact as they begin to modify, create, exchange, and summarize big-picture information. They take this step to make it very clear what action should be taken in upcoming cells of the matrix. A mode of product definition internalization begins. Complexity, conformity, invisibility, speed, and "can do" issues will surface. Some will be resolved; others will be channeled into future work assignments. Most of the time, this is where the product development team will become intimately acquainted with the work done by all their predecessors and begin to understand the possible extent of the job that lies ahead.

The product description reaches a new level of refinement. Simple verbal descriptions make way for sketches, models, and cursory simulations of the final product. The essential visual images of the product begin to form. Team members will make creative use of the tools and materials they have at their disposal to build the visual models of the end product. As the image of the product nears realization, people from research and development, manufacturing, and marketing start to visualize some of the detail that leads to the final product. The first thoughts about work breakdown are initiated. Almost simultaneously, images begin to form of other elements that are needed to create the final product. Manual writers have images of the documents they must create; manufacturing managers start to view the material flow, fabrication, and assembly processes needed to produce the product; R&D engineers begin to see some of the detail behind the concept in terms of hardware and software; marketing engineers see the work they must do to better define the product concept. In short, the team becomes witness to the birth of a new product. It can see some evidence of what the product idea will become, but it also becomes painfully aware of how little it really knows about what lies ahead!

Once an initial product idea has been created to fill a perceived customer want or need, growth in product awareness creates growth in customer and market awareness. The more familiar people become with what the product is or can become, the easier it is for them to visualize how it can be used by the customer. This leads to the identification of potential customers. Opinions will be voiced on which application will sell the most product. Preliminary market segments become visible when it becomes possible to start grouping various customer types. The product developers begin to develop the knowledge base necessary to ask intelligent questions about customers, applications, and markets.

Advances in product and customer knowledge then make it possible to visualize the competitive picture. It becomes a relatively easy task to determine if a product similar to the one under consideration is available for sale. If such a product exists, estimates can be calculated about the sales volume achieved by the competition and projections can be made about the market in which the product sells. Competitive awareness—knowledge regarding the strengths and weaknesses of the competition—adds a salient ingredient to the product's "big picture."

Classification of products as first-of-a-kind, me-too-with-a-twist, derivative, and next-generation helps in explaining certain big-picture characteristics.

Generally speaking, describing me-too-with-a-twist and next-generation products is relatively easy, since prior product descriptions are already in existence. Similarly, knowledge about customer desires is the highest for these two types of products. Knowledge of the competition is most accurate for the next-generation effort; however, with some added work, quality data can be assembled on me-too-with-a-twist competitors. Product development efforts of these types are frequently "pulled" into the market. Knowledge from the market is supplied to the product development process to create a product that meets the requirements set by the market. The "big picture" will be in "vivid color" for these product types.

First-of-a-kind products can, at times, be difficult to describe so that they are universally understood. These new concepts need time to ferment in the minds of those who receive early exposure. Moreover, customers may not be aware of the impact the product could have on their quality of life and generally provide little useful information on the concept. Products of this type are frequently "pushed" to market. In a push development process, knowledge of exactly what the product should be and projections of customer response are generally held within a very small group of people led by product and market visionaries. Product, customer, and competitive information will exist, but it can be expected to be difficult to comprehend and to undergo frequent revision. The "big picture" for this classification is frequently in "black and white" and "out of focus" to many observers.

A derivative product development effort presents its own set of unique challenges. The name *derivative* implies there can be many different products serving many different customers; each offering will have a unique set of competitors. All of this generality creates a lack of specificity. Without specifics, many people get confused. Confusion leads to indecision, which in new product development can create misfortune. The challenge is to say enough about the product's potential for change without saying too much, identify only enough core customers to identify a substantial market potential, and focus on a critical few competitors. Broadly speaking, the "big picture" for derivative products is too big to see through one window. An effectively structured hierarchical representation is usually effective. With skill and some experience, individual derivatives from the family can be transformed to look like next-generation and me-too-with-a-twist products to facilitate understanding and information transfer.

Discussion returns to the MONI example to explore how the display division worked through this cell of the definition matrix (see "Division Business Definition Example" in Appendix).

The "big picture" for MONI's 2001 flat color display (see "Example New Product Definition Sheet [NPDS]") was reviewed and updated. Research and development produced a second working prototype using portions of the new manufacturing process. It became obvious to the development team the form, fit, and function possibilities inherent in the product's design. Manufacturing was able to make progress on defining the problems associated with completing work on the remaining process development. They were also able to identify four other process techniques that required further investigation and study. Marketing's views on product performance were further stimulated by the image quality of the second prototype. They began to see a range of other applications.

The marketing specialists from MONI Display Division worked with marketing specialists from other end-user MONI divisions to prepare a preliminary market research program. Since it was not possible to move the prototype display, internal and external MONI customers were invited to attend a preview of the technology. In exchange for an early view of the advanced display, the customers who attended agreed to be interviewed and complete a questionnaire. This feedback was collected and compared with information within the MONI database. The marketability of the new device was beginning to look extremely attractive. Enough preliminary information was being collected to start formulating serious opinions about the necessary performance, features, price, and timing of the product. Initial sales forecasts of 120,000 units per month were beginning to appear very conservative.

As knowledge of the development effort grew, so did respect for the competition. Some of the problems that were being experienced had been reported in technical circles by individuals from other companies years prior to their observance by MONI engineers. From these data points, the development team began to sense that they were behind some of their competition. Zenith and Photonics Technology appeared to have a substantial lead. Getting the needed information free of charge was no longer possible. It became clear to the research and development and manufacturing team that they needed to recruit a few individuals with advanced technical talent from their competition or buy the knowledge from their competitors. The probability of success with the former seemed to be much higher than with the latter. The sentence "Recruiting technical talent from the competition is desirable" was added to the new product definition sheet.

MONI's work on the "big picture" brought the development team into closer contact with reality. Staff size was still relatively small, but these people knew it could not remain at this level much longer. They began to develop the insights necessary to further define the work that lay ahead. Also, a sense of appreciation for the difficulty of the work they were to embark on began to develop. Key issues were identified that would have to be factored into future work plans to ensure success of the new product development effort.

Does your new product development process give you the opportunity to create the "big picture" early in the formation of a new product's definition? When is your "big picture" a reality? What form is it in?

3.1.3 Product Market Information

Sufficient progress has been made to begin intelligently determining the necessary performance, features, cost, price, and time to market of the product. The cross-functional, horizontal interactions within the project team grow in number and intensify in criticality. It is no longer possible for individual functions to work in isolation without producing a product definition that is too inwardly focused and idealistic. The program manager assumes responsibility for information integration and the resolution of differing opinions between functional groups.

Arriving at the first market information numbers is an iterative process that involves a great deal of thought, discussion, and compromise. Information generation is usually initiated by the marketing group by updating and distributing the new product definition sheet. After seeing the "big picture" and analyzing the market database, marketing engineers are usually quite capable of generating the ideal product definition from the market's perspective. Marketing's insensitivity to the realities of business life in other functions is made visible with this initiating action. All too often research and development can't provide the performance and features that were requested. Manufacturing is frequently aware they cannot produce the product at a cost that will yield a saleable product at marketing's price. Both R&D and manufacturing say "no way" to the proposed timing. This action has served its purpose—the channels of communication are now open and issues demanding resolution have been identified.

R&D, marketing, and manufacturing can now begin the process of understanding the trade-offs that will have to be made to get the project done. Marketing can explain to other team members the performance and features that are needed by the marketplace from the customer's perspective. In so doing, they will get feedback from R&D about the design difficulties associated with their proposed performance and feature mix. Manufacturing will provide feedback on the impact each element has on manufacturing processes and product manufacturing cost. R&D will speak to the group about features and performance that can be designed as a function of available staff and time. Marketing and manufacturing start to develop sensitivity to the challenges faced by R&D. Likewise, manufacturing will explain its position so that R&D and marketing develop the sensitivity to their set of problems. This interaction continues over a period of time until a tentative agreement has been reached on product performance, features, cost, price, and time to market. Once this milestone event has been achieved, it then becomes necessary to take the compromise solution back to the market to test for customer acceptance. The strength and forcefulness of customer feedback will frequently inspire continued iteration of the product definition. All through this highly interactive and dynamic process, the program manager serves as the team facilitator and integrator of the information the team produced.

It is important to note that input from customers is the initiator and terminator of the process just described. As mentioned previously, this input is relatively easy to obtain on next-generation and me-too-with-a-twist development efforts.

It is much more difficult to involve the customer in derivative and first-of-a-kind programs. Within such programs, this is the time to objectively test the ideas held by the visionary technical and market people to ensure that the new product's definition of future work is not over- or underestimated. Sometimes the customer can be of great assistance. The U.S. government and large corporations have groups of people who tell you what to do. They write the specifications of a product they need to perform their mission. Generally, the specification is in the form of a written document that can be actively studied and analyzed. At times, modifications are permitted that tend to enhance discussion, negotiation, and communication with the customer. Once agreement is reached on specification, the program manager can stabilize the product definition. At other times, extracting customer feedback can be a difficult struggle. The challenge is to deliver a product that minimizes development time so that market returns can be generated as soon as is practical from the customer's perspective. Ideally, first-of-a-kind efforts should err on the side of doing too little development and derivative product efforts should err on the side of producing too few products. This suggests that the initial products be utilized as market probes to collect market information rather than having the new product development team spend excess time and energy trying to predict market response without a product.

If the cross-functional new product development team is motivated to use quality functional deployment (QFD) methods, this is now the optimum time in the process to make the attempt. In its most elementary form, QFD means collecting customer-supplied information and then planning the development of the new product around this input. This statement means that (1) product developers have to actually listen to what a customer says and (2) they must use the information to aid in the determination of features, performance, cost, price, and introduction timing of the new product. So, before going off in the QFD direction, stop a moment and ask a vital question: Where and how does the team go about getting customer information?

Obviously, methods will differ for each of the four product classifications. Also, expect the information content to vary between product classification. Figure 3.2 uses the QFD "house of quality" concept to illustrate data collection methods and their effectiveness for each product classification. For a me-too-with-a-twist product, sales and service data on existing products are excellent sources of customer feedback for those already in the business. Similarly, written surveys and focus groups receive high ratings. The reason for using these four primary techniques on "me too" products is to ensure that sample sizes are large enough to maintain statis-

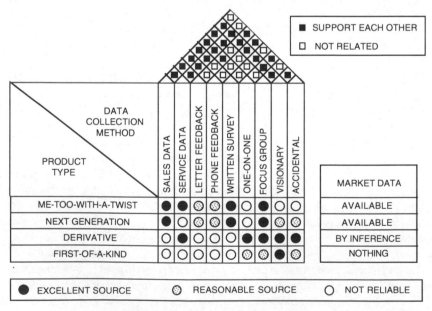

Figure 3.2 Data collection methods.

tical accuracy. In this way, "me too" products can be defined to appeal to as wide a customer base as possible. In this environment, one-on-one customer contact and the ideas supplied by visionaries are usually not very reliable. These inputs become one of many and must be factored in with all of the other responses. Under certain conditions, written and verbal feedback from customers can be put to good use. It takes *statistically accurate* data to operate in a "me too" environment to make winning decisions.

Next-generation market information can be derived from sources similar to those discussed in the preceding paragraph. Sales data are useful for identifying owners of existing products so they can be surveyed on what they would like to see in future products. Written surveys, group interviews, and focus groups are techniques that are sometimes quite effective. Again, as in the "me too" case, strong emphasis is placed on collecting meaningful and statistically accurate information from a representative customer sample—a next-generation product must do what the old one did, only much better and usually for less money out of the customer's pocket. Finding out where areas of give and take are, and their relative sensitivity to one another, becomes a significant determinant in the market success of the next-generation product. Note that the role of the visionary gains in importance. Often, a visionary can see beyond the numbers and make a next-generation product a big success. When this happens, it's typically because the visionary understands the customer's wants and needs better than the customer does. Sometimes it's referred to as walking in the other person's shoes. At Hewlett-Packard, it is referred to as the "next bench syndrome."

The solid-black circles in Figure 3.2 significantly shift in position in the derivative row. Remember, a derivative product is aimed at meeting the needs of a customer in a new market with a product that is quite a bit like one being built for another market. A data-intensive analytic approach is generally not very effective. At best, data points that lie a few standard deviations outside the norm can sometimes be indicative of a potential derivative product opportunity. In fact, customer knowledge for derivative products first appears from the visionary or by accident. The accidental development of markets is a well-documented phenomenon. The way it works is simple. An existing offering gets put to use by an innovative customer who does something new with the product. Someone within your company may witness this and become the internal visionary or the customer may bring it to your attention. In either case, the developers of the new product idea get the early information needed to define customer requirements effectively. Frequently, one-on-one customer contact and skillful utilization of focus groups can be used to solidify generalities into specific quantifiable terms. Market information and the activities in the business column that have transpired are now put to good use in obtaining full support for the derivative product development effort within the organization.

Only one excellent source of customer information exists for the first-of-a-kind classification: the visionary. After all, sales data, service data, written customer feedback, and verbal customer feedback are not possible when concept and product do not exist in the market. When feasible, the new concepts can be taken to select customers for one-on-one sessions. Hopefully, getting something to display to potential customers will lead to a few mistakes that will produce added refinement in the determination of performance, features, cost, price, and timing. As information on these topics begins to stabilize, the focus group environment can sometimes be used to refine information further. In addition to having visionaries, an organization in pursuit of a first-of-a-kind product opportunity must have individuals who work effectively in an environment of high uncertainty.

Knowing the product classification makes it relatively easy to determine which means are apt to work best in terms of understanding the customer's needs. Figure 3.2 occupies a small amount of space but contains a significant amount of information. It now becomes possible to develop and refine the skills of the team to match the techniques that are best suited to bring them into a close working relationship with the customer. If the new product being defined is a "me too," take a course in QFD, learn how to conduct meaningful surveys, and develop data analysis skills. On the other hand, if the new product under development is a derivative or first-of-a-kind, don't spend much time on growth and development in the previously mentioned areas. Instead, learn how to live off your visionaries, build a team of people that work well together with minimal amounts of data, do a lot to encourage the production of mistakes so learning takes place in a safe, controlled environment, and by all means go out and interact with people on a one-on-one basis. Adapt the definition process to the type of product that is under development!

Discussion once again shifts to MONI to explore the work done within this cell of the definition matrix.

Product development team members within MONI were able to resolve the conflicts on performance, features, price, cost, and time to market in a relatively straightforward manner. Their existing databases contained most of the information needed to make the necessary decisions. Where information was lacking, the feedback they obtained from customers in surveys and focus groups provided the required added detail. Most of the team members had worked together in the past and understood the constraints that existed between functions. Performance and feature requirements were demanding, but in view of customer desires and competitive pressure, the goals were set at the minimum level of acceptance by the market. These goals were

$30 \times 18 \times 3$ minimum size	$150 \times 90 \times 5$ maximum size
Color	16-bit resolution
HDTV quality	900×1500 pixels
Computer graphics compatible	95-watts power consumption
Cost: $95	Price: $130

Timing: To market within one year plus or minus 50 percent

Manufacturing was challenged by the cost target of $95 per unit. Marketing was certain that the smallest size display would have to sell for $130 to be competitive. All members of the team agreed that time to market should be two years or less.

This team was now ready to synthesize implementation alternatives that had the potential to achieve the above targets.

With the completion of this cell within the definition matrix, the new product development team's view of the product's definition takes another major step forward. You are encouraged to review the practices used within your organization to set performance, feature, price, cost, and time-to-market targets. How do they compare to those discussed in this section?

3.1.4 Product Alternative Synthesis

With all members of the new product development team armed with market information, it's now time to start exploring designs and methods that can yield a product that meets the market targets. To do so requires a skillful blending of creativity and innovation. Creativity is the thinking process that helps generate ideas about the new product or its method of realization. Innovation is the practical application of the ideas in meeting the needs of the new product development effort more effectively. Alternative synthesis is the mixture of creativity and innovation needed to make the definition phase of the new product development effort a success. Up to this point in the development of the idea, an informal process of synthesis and analysis has been taking place by many associated with the product's development. This cell of the definition matrix draws team attention and focus to the synthesis activity. Now is the time to get all ideas out into the open and begin to explore their potential fully in a team environment. It is time to concentrate on the generation and selection of certain ideas that seem best suited to the challenges ahead. As ideas surface, team members search for effective application methods for them. In effect, individuals and teams are in a brainstorming mode within this cell of the definition matrix.

To give you an idea of how this process works, we'll use an example from a book by Theresa M. Amabile, Ph.D., titled *Growing Up Creative*. She describes the creative process of making soup as follows:

Problem Presentation

You are told to make some soup and you agree that making soup is something you want to do.

Preparation

You gather all the information and resources that will be necessary to solve the problem or do the task when you go through the cupboards and the refrigerator to gather all the ingredients.

Generation of Ideas or Possibilities

You put the ingredients together in whatever way seems best: maybe mostly noodles; maybe vegetables and meat predominate; or perhaps it's a simple stock.

Incubation

You let the soup simmer for a long while; sometimes a new and better flavor emerges.

Validation

You test the ideas generated in the previous two steps. Is it soup? Is it good?

Outcome Assessment

If you like the result, you stop the task that was finished successfully.

If it wasn't quite right, you try again because the outcome was not completed satisfactorily.

If you cannot make it good, you quit and get on to something new.

This example can be related to because of its elegant simplicity and its accurate portrayal of the creation and innovation process. The first step, problem presentation, makes immediate sense. Hopefully, the work done in the preceding nine cells of the definition matrix will make the presentation of the problem to the new product development team this vivid. The business column provides background information. The first three cells, product vision, product big picture, and product market information, present the collective new product development problem to the design team. Preparation, generation of ideas or possibilities, incubation, and validation are delegated to the individual, or team, with responsibility to produce a specific solution to a particular problem associated with the development of the new product. In effect, once the problem is presented, a team of people will be split apart and asked to make a type of "soup" that relates to their area of expertise. As they move forward, they will interact with other members of the team to exchange the needed information on an informal basis. Note that nowhere in these steps are there instructions on how to do this. This open-ended structure is recommended to

1. Develop a new style of thinking based on the problem at hand.
2. Create an appreciation of the complexities involved in the problem.
3. Keep options open as long as possible.
4. Suspend judgment of ideas.
5. Encourage broad thinking.
6. Force remembering past similarities and facilitate forming new associations.
7. Form a fresh perception of the problem.

This is not the time for evaluation, reward, competition, or restriction of choice; to take these actions would severely impair this highly creative phase of the new product development process.

A vivid quotation on the application of the soup example to making performance athletic shoes is given in Reebok's 1989 annual report. Designer David Miller said the following:

> I went to the factory with only drawings on paper. Working with these materials and methods, you have to be there. You have to make decisions right there, without working through a bureaucracy or worrying about second-guessing. I had that running room. When I needed collaboration, I got it. When I needed authority, I got it. That's how the creative process works here.

David's words clearly describe the process of alternative synthesis.

Within the new product development environment, individuals from all functions should be given the freedom and resources to behave as if they were making soup. Creativity and innovation in problem solving are not restricted to research and development. Everyone with a vested interest in the success of the new product can creatively contribute to its realization. The program manager and management team must adjust leadership style to nourish this environment.

From a concurrent engineering perspective, it is possible to observe engineers working in R&D, manufacturing, and marketing begin investigating approaches having the potential to realize performance, features, price, and timing goals detailed in the market information section of the definition matrix. The steps of preparation and generation of ideas or possibilities lie ahead. Many companies place high reliance on prototyping and simulation to generate and view new ideas. Simulation is accomplished using computer models of all major product attributes and the process used in their realization. Electrical engineers can explore various circuit topologies, determine circuit density, and decide on operating speeds. Materials engineers have tools available to simulate the performance of various substances under consideration. Mechanical engineers have the capability to integrate the work of electrical and materials specialists to produce three-dimensional views of the alternatives as they deal with structural and fabrication issues. The tools give these creative and innovative people the opportunity to benefit from each other's specialized areas of expertise.

Working simultaneously with the R&D team are the manufacturing and marketing engineers. Their inputs are fed directly into the simulations to incorporate their needs into the design process while manufacturing and marketing carry out simulations of their own. Manufacturing looks at the alternatives being created in R&D and develops simulations of the fabrication and assembly processes needed to realize the design. Likewise, marketing engineers take the alternatives and start to assess their impact on the market. As these people learn more about the alternatives under consideration, they feed information back to the design team to adjust each idea being simulated.

More ideas will have been considered and explored using simulations than could possibly have been produced in reality. Perpetual thought and study while the ideas are incubating will generally produce a few favorites for continued exploration. At this point, simulation stops and the decision is made to reduce certain high-risk ideas to yet one more higher level of abstraction. This activity is referred to as selective prototyping.

In selective prototyping, those design and process areas of the new product that are judged to be of highest risk are physically realized in some scaled fashion. For example, Burt Rutan's company, Scaled Composites Incorporated, specializes in building low-cost, "proof of concept" test airplanes. It has this to say about itself:

> We provide comprehensive proof-of-concept services at a point in your development cycle that will later allow you to avoid costly changes. From design through flight test, our engineers, technicians, and test pilots serve as a reliable and cost-effective source of data for your development program, be it aircraft, spacecraft or other composite structure. At the completion of our services, you will have the entire picture of static and dynamic flying qualities, both quantitative and qualitative—data unobtainable in the wind tunnel. Yet surprisingly enough, our manned flight test demonstrator programs are cost-competitive with wind tunnel development programs.

Scaled Composites produced an 85-percent size version of the Beechcraft Starship, an 11-seat twin-turboprop high-performance business aircraft, whose flight data proved feasibility. Once feasibility was demonstrated, the data and information obtained from scaled prototyping were used in the design of the final product. For another client, Scaled Composites did something that critics said was impossible and then flew it! Imagine the effect this had on the credibility of the critics. With Rutan's selective prototyping approach, companies can save millions while keeping time to market to a minimum. While Rutan and his company do selective prototyping for a living, many companies have the capability but don't know how to realize its full potential!

From a hardware perspective, key aspects of the design need to be built and tested. This may mean that somewhere between 5 and 10 percent of a new product's hardware design will have to be designed, fabricated, and evaluated. The scaled prototype may not have to work to full specification, but its evaluation must be thorough enough to set the limits on what will be possible in the final design. A recent experience in bringing a high-speed digital-to-analog converter (DAC) to market required the use of this technique. Through simulation, the design team of the DAC was able to predict operational characteristics of the design for an assortment of data, timing, and sampling rates. However, it was not possible to model second-order effects. The only way to determine the amount of undesired signal present along with the desired output signal was to actually build the device and measure the spectral content of its output signal. Measured results on a small number of samples were then used to estimate performance limits for the destination product. Without taking the time to produce the scaled prototype of the DAC, the entire project would have been operating in a complete sea of ignorance. Taking the time to perform this operation took about three months. The knowledge gained shaped the entire future structure of the new product development effort.

Similarly, critical aspects of a software system should be designed, coded, and tested. The process is commonly described as creating throwaway code. Scaled prototyping in software is most often performed to define the human interface, measure the performance of time-sensitive algorithms, clarify hardware and software inter-actions, and observe real-time operating characteristics. Many companies are now beginning to refer to software as "big ware." Features tend to get added at a dizzying pace; graphics adorn most programs; networking capability is frequently implied; interactive performance with other software systems is often a necessity; and oper-ating on various hardware platforms using an array of operating systems is par. The only way to get at major operability issues is with effective utilization of scaled prototypes. Software design teams that know how to apply the technique do so only when they are convinced that pure simulation will not yield accurate answers within a reasonable period of time.

While it is advisable to perform selective prototyping on the product, it is also highly recommended that this method be followed on high-risk areas associated with the process. The product's design will determine over 80 percent of the prod-uct's manufacturing and distribution processes. Hence, selective prototyping done early makes design and process trade-offs very visible. To this end, high-risk pro-cesses that are not well understood should be used to handle their respective por-tions of the product they will be called on to support. If the new product development team thinks it's going to use just-in-time (JIT), design for manufactur-ing (DFM), or some other key process improvement alternative, now is the time to actually try a few things out to set the stage for the future. Only by identifying key process parameters early in the definition phase and communicating them to the cross-functional new product development team can the design of the product be effectively adjusted to meet process requirements.

Selective prototyping of product and process is a painful and arduous experience for many new product development teams—they will find out how accurate their simulations and models performed in the prediction of end-product performance. Once simulations are committed to selective prototype, the incubation process is over. The new product development team enters the validation and outcome assess-ment steps of the creative process on each of the test cases under consideration.

Within this loose structure, the team develops a number of alternative product approaches. Ideally, each approach has information about the user interface (knobs, buttons, keyboard, etc.), machine interface (infrared [IR] remote, computer, etc.), specifications, risks, investments, and timing under development. This information about each product alternative is then supplied to the alternative analysis cell of the definition matrix. Since the process is iterative, it is reasonable to expect feedback from analysis to reignite the synthesis process until convergence is achieved.

3.1.5 Product Alternative Analysis

As product alternatives are synthesized and simulated, and with selected areas of the product and process prototyped, the creative process enters the analysis phase. Here, it becomes possible to validate and assess the results produced in synthesis. The ideas that were synthesized and prototyped are subjected to evaluation and analysis by the cross-functional team. Decisions that help shape the future will be produced.

Will the decisions that get made be based on odds of being right, intuition, analysis and data, or a combination of the three? Or will the organization use a well-known decision style and refuse to make a decision? Making decisions based on odds of being right is a fine style to utilize if the penalty for being wrong is not too severe. You would not want to select product alternative A over product alternative B because on the surface the probability of being right is 0.5. On the other hand, a close-call decision, that can be corrected if wrong, is ideally suited to be made by this method. The intuitive style works well for those cases in which information is limited and the decision maker is a good reader of environment and in possession of some good information. To the casual observer, this decision maker may appear to be working from gut feel, but in reality is making a decision based on full sensory stimulation. On the other hand, an intuitive style that ignores environment and information is not likely to produce quality results. Next, the analytic decision style emerges. Here, the decision maker relies on information to initiate action. This style is effective when the assembled information is accurate, complete, and does not change very rapidly. When this is not the case, the analytic-based decision maker is in trouble. What may look like a combination of styles is fine if the person wielding the power knows how to make informed decisions and understands the strengths of one technique over another. Individuals with the combination talents are usually in small numbers within any organization, but these skills can be developed.

The previously described styles are listed below to show when they are likely to work the best:

	Information Quantity	Information Quality	Error Tolerance	Speed
Odds-based	Low	Low	High	Fast
Intuitive	Low	High	Low	Fast
Analytic	High	High	Low	Slow
Combination	Medium	High	Low	Medium

An odds-based style should be utilized when information quantity and quality are low and error tolerance is high. Think of it as the style to use when you can change your mind a lot and get away with it. The analytic style is the exact opposite. It should be used only when information quantity and quality are high and the resulting decision has little or no margin for error. A typical situation in new product development that demands an analytic style is when research and development must decide on actual performance specifications of a product. Between the two lie the intuitive and combination styles. A combination approach is preferred over a pure intuitive method, but it can only be utilized when information quantity and quality exist in greater proportion. Typically, an intuitive decision can be made ahead of a combination decision. In business, this sometimes is the determinant of success when decisions are viewed as a collection of serial events.

Individuals doing alternative analysis must be sensitive to decision style to keep the process moving at an aggressive forward pace. Also, knowing when to apply what style helps in making adjustment to the analysis cell if the situation appears to be stalled.

Discussion now leaves decision style and returns to describing actions that take place in the product analysis cell.

Research and development will be interested in the determination of preliminary performance figures for each alternative under consideration. It's quite likely that the R&D team will want to stress test the scaled prototypes under a variety of environmental extremes to determine temperature, altitude, and mechanical effects. Engineers will measure performance under a variety of operating conditions and compare actual performance numbers to those predicted by the simulations. A conscientious design engineering team will systematically go about the process of reconciliation.

Manufacturing participates in the creation of the scaled prototypes by building them using existing manufacturing processes and developing scaled prototypes of new processes in areas where manufacturing capabilities do not exist. In this manner, manufacturing gets the opportunity to evaluate their part of the synthesis – analysis experiment fully. A numerical comparison between simulations and actual results is now possible. The comparison and reconciliation process points to areas of strength and weakness in manufacturing. This analysis will indicate the areas where continued process development is required so that the manufacturing engineering team can focus its efforts. Alternatively, the experiments will highlight design issues, which when addressed by appropriate adjustments in design, will better align product with process.

Marketing becomes an active observer of the measured results as it becomes more familiar with the product alternatives from subjective and objective perspectives. As marketing engineers review performance data and formulate impressions from simulation and scalable prototyping activities, they are able to refresh the information in their sales models and update thoughts on product promotion. This action, as was the case for research and development and manufacturing, gives marketing the opportunity to reconcile differences in opinion and data to help focus on the selection of the final alternative.

Information from analysis is fed back to synthesis in an iterative manner to converge on the final results for each idea being assessed. With each pass through the synthesis–analysis loop, numerical and subjective evaluation criteria begin to stabilize across functional areas. It becomes possible to gain numerical insight about the sales potential (mature sales volume), profit potential (net profit percentage, total annual net profit, etc.), development resources requirements (R&D engineers, manufacturing engineers, marketing engineers, facilities, support, etc.), mature manufacturing costs, risks associated with continued development, and critical issues that lie ahead to arrive at an internal rate of return (IRR) for each development alternative. Decision makers will utilize this data to select the alternative that best meets present perceived business needs. Previous work on the definition matrix now receives a key test by placing decision-making criteria directly in front of the new product development team. The process of making relative decisions continues, but it is firmly guided to its termination point. Key people on the team will be actively involved in helping decide which path to follow. At times, the selection of a final development alternative becomes a close call. Performance numbers might be close, manufacturing cost differentials may be small, and returns might be comparable, but the clear choice will emerge. Once the final decision has been made, work on finalizing the product definition and detailed planning of future work can begin—the new product development definition phase is almost complete.

As might be expected, first-of-a-kind and next-generation product development efforts will spend more time in analysis than me-too-with-a-twist and derivative product development endeavors. The former new product development teams are likely to deal with far more unsolved design and development problems than the latter. Additionally, first-of-a-kind and next-generation product development efforts are likely to experience far more returns to the synthesis cell of the definition matrix owing to the fact that these efforts are struggling with more complex issues. Me-too-with-a-twist efforts can be expected to move through the analysis cell if the development effort has been scoped to attack only one or two critical variables in the continued evolution of the iterative product. As mentioned before, the strategy of controlling the scope of change is an important ingredient in holding down time-to-market numbers on me-too-with-a-twist products. Under these conditions, it is quite likely that the analysis of alternatives will yield a selection without return to the synthesis cell. Derivative product development efforts are likely to progress through analysis in a time frame that is proportional to the complexity of product change; that is, greater complexity will require greater time and may cause repeated return to the synthesis cell.

Discussion again returns to MONI.

MONI's new product development team produced three alternatives for detailed analysis. Most of the electronics between the items under evaluation were the same, while each item differed in display materials and display fabrication processes. Of course, differing materials and processes produced slightly different mechanical configurations in the evaluation prototypes. To add statistical meaning to the experiment, manufacturing, led by a key materials process engineer recruited from NEC, elected to produce a valid number of display samples with each new process to assess process variability accurately.

As planned, each of the functions began to measure the performance of the prototypes to collect objective data and form subjective opinions. The statistical experiments in manufacturing generated significant amounts of relevant data. Upon integrating all of the cross-functional information, it became obvious that one manufacturing process dominated over the others in terms of yield and that the prototype built using this process produced acceptable visual performance. The selection of the best alternative in this case was being made on manufacturability first, performance second, and time third. Other factors remained in the background. Also in the background was a patent infringement risk with a leading competitor.

The MONI new product development team, while nowhere near the completion of the project, had an unobstructed view of the work that remained to be completed and had sufficient information available to help others clearly see the future. All that remained were the tasks of collecting and presenting the information that had been assembled to complete the project's definition. Please refer to the product definition sheet ("Example New Product Definition Sheet [NPDS]") within the Appendix to refresh your view on where this project is headed.

3.1.6   Product Definition Output

The end of the definition phase is near. Eleven cells of the definition matrix are completed. A business definition helped pave the way to effectively start and work on the production of a product definition. Throughout the product definition process, team members worked horizontally and vertically within the organization. Research and development, marketing, and manufacturing became acutely aware of the total scope of the development effort and learned how their work would contribute to the new product's success. As definition work was completed, a history of accomplishment was produced. The new product development team has now generated and assembled sufficient information to begin using it in subsequent activities.

The mere fact that a wealth of information about the new product development effort exists is a major step forward, but something is still missing—data from all preceding 11 cells must be refined and focused on just this one particular new product development effort. If all this information had physical form, the process to this point would have produced an office littered floor to ceiling with models, prototypes, charts, and stacks of paper. In our metaphor, a rush of air, if strong enough, would delay the project for many months. The office we describe has pathways leading from one key piece of development information to the next. Dust has settled on most of the early models. Only a few people know how to separate the quality items from the chaff; there may be others that disagree with these individuals. Clearly, time needs to be set aside to collect, reduce, simplify, organize, distribute, and approve the key information that will be used in the future. The small team that did the definition work must collect its thoughts and prepare the documentation for future participants.

Visualization of the final definition work that needs to be performed is essential for continued team success. The model that provides this visual structure, and the framework for planning and execution as well, is the new product development process (NPDP) matrix. The full NPDP matrix is given in the Appendix. Take the time to acquaint yourself with its structure. It has work functions arranged along the horizontal (program management, customer documentation, customer support, sales, etc.) and time along the vertical. Work that needs to be performed as a function of project phase is listed at the intersection of each row and column.

In the process of creating the matrix, many corporate structures were studied and compared. It was found that 12 functional classifications existed, in some form or other, within just about all of the organizations evluated. The functions identified were

Function	Key NPDP Responsibilities Identified
Program management	Project/program definition, planning, and execution; leadership and management; horizontal and vertical communications
Customer documentation	User, training, and service manuals.
Customer support	Service and support strategy; service and support systems; application training; service training; field repair; telephone application and service support
Sales	Sales forecasts; sales-related training and promotion materials; product sales; order processing
Product marketing	Promotion strategy; advertising; promotion literature; product descriptions; market research
Quality assurance	Regulatory compliance; product reliability; product safety; process statistics
Manufacturing	Raw materials supply and flow; vendor relations; scheduling; parts fabrication; product assembly; product test; distribution
Personnel	Recruitment; employee development; labor–management relations; wages and benefits; outplacement
Finance	Operations accounting; cost accounting; financial analysis; financial reporting
Systems engineering	System specification, integration, and evaluation
Hardware engineering	New product hardware design, development, and test; system integration and evaluation support
Software engineering	New product software design, code, and test; system integration and evaluation support

Review of this information makes it clear that a number of people, doing many different but related things, need to generate, exchange, and modify information to contribute effectively to the development of the new product. The NPDP matrix model serves to bring visual order to a highly interactive and complex system. One level removed from this matrix is a functional management layer. In a typical organization, systems engineering, hardware engineering, software engineering, and customer documentation will report to the research and development functional manager; customer support, sales, and product marketing are led by the marketing functional manager; and manufacturing, personnel, quality assurance, and finance are, most often, headed by their own functional managers. Within these organizations, only the program manager has responsibility for working across all disciplines while working with the management team on a horizontal and vertical basis.

At this point, a review of Section 1.4, "Organizational Models," is recommended for a full understanding of how the NPDP matrix fits within the overall framework of an organization. This is also a good time to sketch a NPDP matrix for your organization. You'll be able to determine where your organization differs from the model described as you develop a better understanding of the matrix modeling concept.

With the NPDP matrix in hand, each functional team is responsible for the production of output listed in its respective cell along the definition row. To set the example for what takes place in all cells, we'll discuss the work that is performed within the program management matrix rectangle. The deliverables requested within the program management cell of the definition phase are

1. Production of a project/program definition document.
2. Measured prototype benchmark performance.
3. Production of a draft project plan:

   Estimated program time to market (TTM) based on documented complexity and productivity figures.

   Estimate of development costs.

   Estimate of resource requirements.
4. Presentation materials to secure organization go-ahead and resource support to enter the planning phase.

The definition document is compiled over time as each cell of the definition matrix is completed. It consists of an information structure that defines the external operating characteristics of the product, and it is frequently referred to as the external reference specification (ERS). The primary function of the ERS is to communicate what the product is (form, fit, and function) and how it is to be used (human interface and controls, machine interface and response, electrical interface specifications, mechanical interface specifications, etc.) so that others can develop a fully integrated view of the product. Development of this document is an iterative process that starts out slowly and grows in complexity as the definition effort moves down the product column of the definition matrix. As the definition phase reaches its point of termination, the ERS becomes the vehicle for communicating detailed project information *to* the design and development team; it represents the best effort of the program manager to define the product that is to be built. It is a symbol of how well the program manager was able to reduce all the work that took place during the definition phase into concise, specific, measurable terms. In the future, measurement of change to this document will serve as an indicator of definition stability. An expanded discussion on the ERS follows.

From an overview perspective, the first section of the ERS might contain information extracted from the NPDS. Likely elements to utilize are the internal purpose, external purpose, resource summary, sales projections, cost estimates, and schedule figures. In support of this information, the product description, which is limited to a few lines of text on the NPDS, is expanded. Next, thought and work should go into the production of visual images of the product. Views of the outside, front, back, top, bottom, sides, and sometimes the inside are appropriate. Where justified, pictures of the user interface and machine interface are produced. Quality of the overview section of the ERS is judged by its simplicity and ease of understanding by members of the new product development team. Done efficiently and effectively, the overview section typically consumes less than three pages of text and graphics.

In the "form" section of the external reference specification, consideration is given to supplying information on the product's dimensions, weight, color, and configurations. This general level of information is followed by supporting specific information on location and attachment of human interface input and output devices. Likewise, the same level of attention is given to the location and attachment of the machine interface input/output devices. Utilization of the data in this section makes it possible for members of the new product development team to understand fully the external physical details of the product from a visual and numerical perspective. The type and location of screens, microphones, buttons, panels, slots, and connectors are fully specified. An overview of what goes into or comes out of the artifacts is assembled and documented. Detail and complexity expand in this section. It is not unusual to have certain drawings enlarged to C or D size and to have a number of views for each of the human and machine interfaces.

Descriptive effort shifts to providing detailed information on human and machine stimulus and response parameters. In effect, the "function" section of the ERS describes system properties of the product. Where human-input devices are concerned, the utilization and range of expected input are specified. For valid input, expected product response is then specified. Also, attention should be paid to the system's response when input is out of the expected range. Where appropriate, performance specifications relating input to output are established. Some products respond to touch, but many new products are becoming sensitive to sight, sound, and motion. Consideration must be given to each allowable input mechanism. In addition to describing the range of human and machine interaction, many products also require the complete determination of stimulus/response between machines. In this case, signals from an external machine will activate the product. Its response to various stimuli must be specified. Most often, signal and mechanical interfaces between machines relate to certain standards established within the communication and computation industries. The "function" section of the ERS can become rather detailed and extensive. Most of the required information is classified as system-level and should be given full attention by those fully responsible for the system attributes of the product. When done effectively, members of the new product development team will know how the product is expected to respond to the range of anticipated human and machine inputs.

Lastly, the external reference specification provides the new product development team with related information about the product that enhances cross-functional communications. Marketing may want to add a pro forma data sheet to help the team members understand how it views the product. Manufacturing may want to provide detail on the processes it expects to utilize in the production and distribution of the product. Sales may add a section on expected product order configurations and its expected announcement and introduction structure. Detail after detail that the cross-functional team has thought about and somewhat understands should go into the document. Naturally, information should be appropriately filtered and reduced to ensure its acceptance and broad utilization. The last page of the ERS should summarize the numerical specifications of the product.

Production of the ERS is a collective team effort. Systems engineering and the program management staff can think of themselves as editors of the document. A new product development team producing its first ERS may have to allot a few weeks specifically for the production of the document. A group that has had a few members go through the process on a previous effort can produce the ERS simultaneously with the work of the definition phase. At worst, only a few days will be required for final editing and production.

At the same time it develops the ERS, the new product development team is integrating simulation and scaled prototype data. Rarely does a full prototype of a final product exist during the definition phase. As described earlier, decisions are made to selectively prototype only those sections of the product that appear to have the highest risk. Other elements of the product are evaluated by modeling and simulation. The program manager and the team must wade through a sea of data that was produced by modeling, simulations, and focused prototype experimentation. This data must convince the program manager that future risks are within acceptable personal limits. A summarization of this data must also persuade the development and management people that the risks are within their acceptable bounds. It's not likely that a project would move out of the definition phase without some effort being devoted to the topics of problem understanding and risk management. Selective prototyping and effective simulation can be used to bring both into control.

Next comes the task of producing a draft project plan. Rather than suggest that the program manager make the best guess possible under prevailing circumstances, a technique often employed, I'm going to offer an alternative approach that has demonstrated its effectiveness on a number of programs. The technique uses time to market (TTM) as the dependent variable and work complexity, worker productivity, and work group size as independent variables. To use the technique, the program manager solicits TTM information from each team contributing to the development effort and asks for their supporting data. In effect, all team members become a part of the estimation process. When the project enters the planning phase, the importance of using this estimation process will become evident.

How the approach is used to make hardware engineering and software engineering estimates is thoroughly discussed below. Attention will then shift to the use of the technique by other groups identified on the new product development matrix model.

Attempting to predict the calendar time duration (TTM) of hardware and software projects is not for the weak at heart. Most of the time, an educated guess is made and the planning activity is started. This is just not good practice. To do justice to the planning phase, it must be entered into with reasonable definitions and expectations. It's similar to saying you know the right answer to within about 25 to 50 percent. Results from planning can then be fed back to modify the definition and converge on the correct answer to the time-to-market question.

Is there a relationship that can help solve this dilemma?

Fortunately, the answer is yes! Over the years, I have had the opportunity to manage and study a number of software and hardware projects/programs. In so doing, data was recorded on staff levels, person-months of effort, job complexities, elapsed calendar time, and other relevant project information data. Moreover, others in the work environment were collecting data. This information was exchanged, discussed, and reviewed. For the most part, some conclusions were drawn, but little practical use was made of them in helping define and plan future projects.

In the course of writing my first book, I went back and reviewed the information one more time to see if anything useful could be extracted. This time around, key pieces fell into place to make sense of the data. First, I observed that work output seemed to follow a linear relationship when viewed from a historical perspective. After the work was done, what was the average daily productivity that led to the production of the resultant work output? I looked for measures that related directly to final output. All other activities that led to the production of the output had to be contained in the measurement. The goal was to develop an estimation method and not to develop in-process monitoring techniques. For software, lines of code per day were the right metric; for hardware, components per day. Data showed variance from person to person by some amount, but for any individual, the data suggest that output is somewhat the same over time when similar work is assigned. Second, project durations seemed to relate to the complexity of the work (total number of components for hardware and total number of lines of code for software) and the number of people that were assigned to execute the projects. Third, smaller work groups usually had a higher productivity than larger work groups. I put some points to paper, drew some lines, and repeated the process for a number of different hardware and software projects. Amazingly, a simple pattern surfaced that could easily be represented mathematically. Two key equations are given:

$$\text{TTM}_{h/w} = \left(\frac{N_c}{N_e}\right) \left(\frac{1}{P_h}\right) \left(\frac{1}{260}\right) \quad \text{(hardware)}$$

where

$$
\begin{aligned}
\text{TTM}_{h/w} &= \text{calendar time duration for a hardware project} \\
N_c &= \text{number of equivalent components} \\
N_e &= \text{number of engineers assigned to the project} \\
P_h &= \text{hardware productivity (components per day)} \\
260 &= \text{number of annual working days}
\end{aligned}
$$

$$\text{TTM}_{s/w} = \left(\frac{N_l}{N_e}\right) \left(\frac{1}{P_s}\right) \left(\frac{1}{260}\right) \quad \text{(software)}$$

where

$\text{TTM}_{s/w}$ = calendar time duration for a software project
N_l = number of lines of uncommented source code
N_e = number of engineers assigned to the project
P_s = software productivity (lines per day)
260 = number of annual working days

Figures 3.3 and 3.4 present graphs pertaining to the hardware equation, and Figures 3.5 and 3.6, for the software equation.

Figure 3.3 holds N_e (number of hardware engineers) at a constant value of 1 while showing variability for a number of values of P_h. An average hardware engineer in my group performed within the 0.4 to 0.6 component-per-day region. This is not to say that there are no people that perform outside of this region. In particular, I have a few to single out that can outperform their peer group by factors that range from 2 to 6. The curves do tell you, once the component complexity (part count) is set and an engineer with a particular productivity has been assigned, your TTM has been established for a portion of the project. Adjustment of work distribution, engineers, and the task network becomes a major challenge for the hardware engineering manager to arrive at, and then deliver, a TTM objective. As you are well aware, staffing decisions can make or break some projects. This simple model can be used to get work appropriately assigned down to the individual hardware engineer. Remember that lower productivity rates apply as individuals are placed into progressively larger work groups. In effect, increasing team size mandates that estimates be made by utilizing curves at a lower productivity factor.

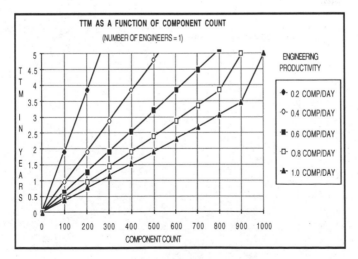

Figure 3.3   Hardware TTM by person.

Care must be taken to get a determination of the average productivity levels in your environment. They are frequently easy to reconstruct by collecting person-month data on projects that have been completed and counting up the component complexity in the resulting product design. These numbers can then be adjusted to arrive at an average hardware productivity figure. There is no need to associate productivity figures with specific individuals. Average productivity rates are usually all that are required during the definition phase. Detail productivity assignment and analysis will be done in the planning phase. You can use the information now to get at a calibration point to help you decide how many hardware engineers your project will need. In the planning phase, you will be able to look at the productivity being asked of each engineer to makecertain it is within a reasonable range.

Figure 3.4 maintains the complexity (N_c) constant at 1000 components and solves the equation for various values of P_h. This component count was selected to make scaling the curves to more complex projects a simple exercise. To arrive at a TTM estimate requires that you use average productivity information and engineering count. Likewise, TTM and productivity can be used to get an estimate of the number of engineers that will be needed by the project. Please note some very important points about the family of curves. One, productivity has high leverage! On a two-engineer hardware project, a swing from 1.5 to 0.5 can move TTM from one to four years. This should help you understand why it is extremely important to hire the right people, provide them with the right tools, maintain a quality working environment, provide a well-structured assignment, and set up an incentive system that keeps these people motivated. Two, at a constant productivity, it is necessary to double staff to cut TTM in half. This can come as quite a shock to a number of managers not used to dealing with this concept, but it goes straight at why adding one or two people to a large project doesn't get much results. A 10-person team

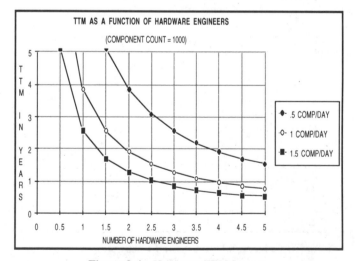

Figure 3.4   Hardware TTM by team.

won't cut its TTM by a factor of 2 until its size is doubled to 20. However, increasing staff by this amount places you at risk of lowering productivity and therefore of not seeing its full benefit. Lastly, the curve makes it clear that arriving at a TTM estimate is a delicate balance of staffing, productivity, and assignment complexity. Any manager with some experience will recognize that this data points in the right direction.

Figure 3.5 transports us from hardware to software. Productivity in this environment is well studied and quite well documented. The industry, as reported by Capers Jones and Frederick Brooks in their publications *Programming Productivity: Issues for the Eighties* and *The Mythical Man-Month,* respectively, averages a productivity of about 10 lines per day. I have personally observed productivity ranges from 6 to 50 lines per day that depended heavily on the individual and the assignment. In general, real-time system software development productivity was lower than productivity on other assignments. The following chart holds N_e (number of software engineers) at a constant value of 1 while showing variability for a number of values of P_h. As in the hardware case, the curves tell you, once the complexity (line count) is set and an engineer with a particular productivity has been assigned, your TTM has been established for a portion of the project. The software engineering manager faces the challenge of adjusting work distribution, engineers, and task network to arrive at, and then deliver, the software TTM objective.

Figure 3.6 maintains the complexity (N_c) constant at 10,000 lines and solves the equation for various values of P_s. This graph points to similar information as the hardware graph. The dependency of TTM on staffing, productivity, and assignment complexity takes on a like meaning. To arrive at a TTM estimate requires that you use average productivity information and engineering count. On the other hand, TTM and productivity can be used to get an estimate of the number of software engineers that will be needed by the project. Again, note that productivity has high

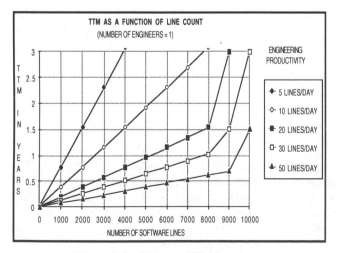

Figure 3.5   Software TTM by person.

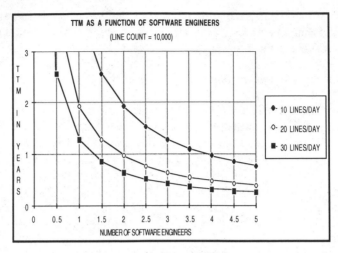

Figure 3.6   Software TTM by team.

leverage! A two-engineer software project swing from 30 to 10 can drive TTM from 0.75 to 2 years. This helps explain why the industry has placed so much emphasis on software productivity. Also, the need to double staff to cut TTM in half remains a part of this model. Amazingly, there has been a greater acceptance of this phenomena in the software teams I've managed than with those who work on hardware and systems.

We will now move on to review a few examples that I've had the opportunity to observe and apply some of the data from these projects to these newly developed models. For the first example, consider a hardware project that was first estimated to have the following values:

$$N_c = 2500 \text{ components}$$
$$N_e = 6$$
$$P_h = 0.5 \text{ components per day}$$

This results in

$$\text{TTM}_{h/w} = 3.2 \text{ years}$$

As the project progressed, the component count was found to grow by 10 percent. Budget restrictions kept the organization from hiring the sixth engineer. Construction work in the hardware engineering area reduced productivity to 0.4 components per day. On the surface, these changes would seem to be reasonable and would not be expected to have a big effect on time to market (TTM). Are they reasonable? Try solving with the new values:

$$N_c = 2750$$
$$N_e = 5$$
$$P_h = 0.4 \text{ components per day}$$

This results in

$$\text{TTM}_{h/w} = 5.29 \text{ years}$$

TTM has just been blown away! By permitting things that, taken individually, seem to be tolerable, we have a slip of 2.1 years. For all practical purposes, this project is lost. Its market window has been eaten by the competition. The example serves to illustrate that control of TTM is a game of inches and not yards. Little things done right make you a winner; little things done wrong can lead the whole organization to disaster.

Now let's look at an organization that follows the suggested process. In a project to code a math accelerator card, the following estimates were derived from experienced software developers:

$$N_l = 50,000 \text{ lines}$$
$$N_e = 9 \text{ engineers}$$
$$P_s = 10 \text{ lines per day}$$

resulting in

$$\text{TTM}_{s/w} = 2.14 \text{ years}$$

During the planning phase, it was determined that P_s would run at 15 lines per day and that code complexity would grow to 60,000 lines. Not willing to bet TTM on the new productivity estimate, the program manager added one additional engineer to the effort. Plans were adjusted and the project was executed with the following parameters:

$$N_l = 60,000 \text{ lines}$$
$$N_e = 10 \text{ engineers}$$
$$P_s = 12 \text{ lines per day}$$

This resulted in

$$\text{TTM}_{s/w} = 1.92 \text{ years}$$

Complexity grew by 10,000 lines. Productivity only improved by 20 percent. The addition of one additional resource was sufficient to keep the project on track and to meet its two-year time-to-market goal.

As a final example, a product development effort based on optimism, "can do," and a major error in staffing is considered.

A few years ago, I had a number of discussions with associates involved with a start-up activity. The people involved spoke enthusiastically about their future. Their plans to develop the greatest of all products and the fortunes they would accumulate were the subject of many hours of social conversation. The entrepreneurial forces were at work! During the discussions, I inwardly felt that something was wrong with the situation, but could not identify why.

This group of individuals and their product development team worked long days, plus extended weeks, for a period of almost three years. In the first years, there was excitement and optimism. A success or two would appear every so often to keep motivation up. As time went on, this spirit began to fade . A touch of reality set in. This team began to understand that there were just too few of them to get the work done. Attempts were made to get more funding so that staff levels could be expanded. These efforts were not successful. The next action was to modify the product definition. It resulted in a product that was below customer expectations. The startup failed.

In subsequent discussions with key individuals involved, it was established that hardware complexity N_c was about 7800 components and software complexity N_l was about 110,000 lines. Funding and management decisions limited the number of hardware engineers to six and the number of software engineers to eight. Giving credit (maybe excessive) to the start-up work environment to put hardware productivity P_h at one component per day and software productivity P_s at 20 lines per day resulted in

$$\text{TTM}_{h/w} = 5.1 \text{ years}$$
$$\text{TTM}_{s/w} = 2.6 \text{ years}$$

The software effort was just about right to hit a three-year time-to-market goal. The addition of six hardware engineers would have brought $\text{TTM}_{h/w}$ into alignment with $\text{TTM}_{s/w}$. The incremental cost to add this staff would have been less than 7 percent of the funds that were expended over the limited lifetime of this company! Again, application of the relationships could have helped this company get the definition of the product into alignment with the realities of their business situation.

Obviously, to apply this technique, you will need to determine the correct productivity rate for your organization (I have observed variations in P_h from 0.2 to 1.6 and in P_s from 6 to 50), establish a method to estimate hardware and software complexity, and prove (disprove?) the validity of the equations in the new environment.

To move the estimation technique throughout the new product development environment, managers of other teams need to develop productivity measures for their work groups that operate outside of hardware and software engineering.

The productivity of each group is likely to be measured differently. The first step is to identify the main activity undertaken by the group and focus on the resultant output. Every activity that leads to the final output should be a subset of the overall productivity measure selected. Working with a number of measures on an initial basis tends to make it easier to isolate the key one or two measures that are most relevant to a particular function contributing to the realization of the new product. To stimulate thought on possible productivity measures a given function might use, the following list is presented:

Function	Suggested Productivity Measures
Program management	ERS pages per day; NPDP matrix elements per day; NPDP network elements per day; revisions per day; problems found per day; problems solved per day
Customer documentation	Pages per day; illustrations per day
Customer support	Pages per day; illustrations per day; parts per day; customers per day
Sales	Sales forecast data items per day; training and promotion pages per day; product sales per day; orders processed per day
Product marketing	Advertising pages per day; promotion literature pages per day; product description pages per day; market research data items per day
Quality assurance	Compliance tests per day; product reliability tests per day; product safety tests per day
Manufacturing	Raw materials supply and flow rates; vendor response rates; parts fabricated per hour; products assembled per hour; product tests per hour; products packaged per hour
Personnel	College hires per month; experienced hires per month; employees trained per month; employees terminated per month
Finance	Operations account pages per month; cost account pages per month; financial analysis pages per month; financial reports per month
Systems engineering	System specification pages per month; system analysis pages per month; system test data pages per month; revision pages per month
Hardware engineering	Analog components per day; digital components per day
Software engineering	Uncommented lines per day

Knowing the productivity of each of the work groups and estimating the complexity of their work allows for the prediction of the overall project/program time to market. Estimation of the program's TTM is delegated to those responsible for the future planning and execution of work which helps prepare them for these activities. It is likely that the overall estimate will be based on the input from dozens of estimators each with expertise in a given area. The program manager is the first to see the feedback from the various groups and the supporting data they used to derive their estimates. Having collected the dependent variable and the independent variables that led to the generation of the individual TTM estimates places the program manager in an excellent position to interact with those doing the number generation. He or she then begins a process of interaction to internalize, challenge, modify, and eventually accept the input received. In a short period of time, an overall preliminary plan will be synthesized that links the information supplied by the functional estimators to arrive at an overall TTM for the new product development effort. If mistakes were made in the estimation process, their effects will be averaged by the number of estimates utilized in the prediction of the overall program TTM. Accordingly, confidence in the result should build as the number of inputs increases. The process just described is no longer an exercise based on faith and luck—it is an exercise grounded in fact and solid engineering judgment.

As definition work nears completion, members of the team can devote some time to reviewing the new product development process matrix shown in the Appendix. Twelve cells, one for each primary function, have a checklist to scan to determine if their work output for the definition phase is truly complete. For example, examination of the manufacturing cell reveals the following checklist:

Design for manufacturing process documents
    Information requirements
    Cycle times
    Wage and overhead rates
    Volume dependencies
Process cost models
    Software
    Printed circuits
    Product assembly
    Product test
    Integrated circuits
    Material acquisition
    Material handling
    Material distribution
    Material fabrication
    Hybrids
Concise written description of other deliverable items
    Engineering services
    Material services
TTM_{mfg} estimate
Estimate of full staff needs
Estimate of execution cost
Estimate of complexity

The manufacturing team is asked if needed design for manufacturing (DFM) information, such as cycle times, volume dependencies, wage and overhead rates, and general information requirements about each manufacturing process, has been distributed to the design team. If it has done this, it then makes it possible to optimize the design for each documented manufacturing process. Likewise, the second item on the list (process cost models) is provided so that the design team can use process cost models to optimize the design to minimize manufacturing costs. The third item (concise written description of other deliverable items) links manufacturing to other functional groups through the output that will be produced. These three steps, initiated by manufacturing, bring product design and manufacturing processes into harmony very early in the development of the new product. The last four items on the checklist deal with the production of quality estimates by manufacturing. Has the manufacturing team estimated the complexity of the work it expects to perform? Knowing this number and the average productivity of the team, how many people are required to achieve the team's time-to-market estimate? Based on person-months of effort, how much funding will manufacturing require?

Each group along the horizontal axis of the new product development process matrix is asked to review its respective cell and provide the identified information to the program manager or the appropriate work group. Once complete, integrated, and accepted, the information leads to the final step in the definition phase.

Armed with an external reference specification, measured performance on high-risk areas of the product, and a completed NPDP matrix checklist, the program manager makes a series of management presentations to secure the go-ahead to have the project enter the planning phase.

Attention in meetings of this type will usually focus on financial and developmental risk issues. The program manager must be prepared to present new product development staff requirements as well as other detailed developmental expenses. Staff figures and expenses can be used to develop the estimated investment cash flow for the new product development venture. Estimated sales and production cost figures, combined with investment financial information, are used to compute the internal rate of return (IRR) for the project and the project's break-even time (BET). Both of these calculations can provide insight on the desirability of going ahead with the effort. If these bottom-line numbers and the methods used to produce them meet expectations, then it is likely that the sponsoring management team will want information on developmental risks that lie ahead. The discipline of following the definition matrix and completing the entries within the NPDP matrix forces the identification of many risk issues. Thus, the program manager has a stable base from which to address these management concerns. Most likely, a well-prepared program manager will receive the go-ahead order from an enthusiastic management team.

We'll now return to briefly explore the work that was performed by MONI's program manager to prepare to complete the definition phase of the 2001 flat color display project.

The external reference specification for the display was in continuous development throughout the definition phase of the project. Portions of the document were developed as each cell of the definition matrix was completed. As each element of information surfaced, it was inserted into a product definition template that was developed by the program manager from the definition matrix. This template made organizing the information an easy task. The document fully described the data and power electrical interfaces to the display, the mechanical interface to the display, and the full electrical, mechanical, and environmental performance specifications. The ERS for the product was a 19-page document that had been reviewed and accepted by each functional area interested in its information content. Upon exit from the definition phase, the document would go on a strict revision control process to make change in ERS information content a very visible act throughout the display division.

Prototype performance spoke for itself. A specially prepared high-resolution source was used to generate image data directly from 70-mm film. The source had a buffer memory that would allow storing up to 10 minutes of continuous playback. A survey was sent out to determine the favorite movies of the MONI display division management team. From the survey information, a sequence of segments was formed from the most popular selections. To add further excitement to the prototype display, the engineering team had available laser disk, video camera, and 3-D computer source drivers that could be connected to the display under remote radio-frequency (RF) control. The future of MONI's display technology hung on the wall in its display division's development laboratory for team and management review.

Development estimates were made using the time-to-market technique previously described. These estimates forecast TTM to be 26 months. The total development cost was predicted to be $4.5 million. Adding marketing sales information produced a break-even time (BET) of six months and an internal rate of return (IRR) of 23 percent. Key risk and concern areas were identified.

A short series of review meetings were held to secure go-ahead approval to continue the project and enter the planning phase. At the last such meeting, the management and development team reached the same obvious conclusion: continue!

The new product development team can now precisely locate the completion of the definition phase in time—the phase ends when the new product development process (NPDP) matrix is completed by the project team and when management acts on the summarized results. A decision to move forward results in the commitment of capital and resources to produce a detailed plan of the execution phase based on the product's definition.

3.2 PROBLEM ANALYSIS

The work that has been completed by the development team to this point has had a major impact on the team's future. Recall the list of new product development process problems that were identified within section 1.3. Producing a quality product definition has a major positive impact on 40 percent of the items on this list. The elimination of these definition-related obstacles has a significant influence on team performance and is certain to reduce delay in subsequent phases of the new product development process. Subsequent discussion focuses on those problem areas that can be corrected by improving definition phase techniques.

Poor (Changing) Product Definition. The development of a quality product definition is not a single-step process done by an individual in a vacuum. It is a horizontal and vertical team effort that can affect just about everyone involved in the new product development process. The definition matrix establishes a recommended path to follow to generate a stable product definition. Six of the matrix cells detail the business definition steps; the remaining six cells guide the product definition effort. Final product definition requirements are established by the new product development process (NPDP) matrix.

A good product definition is an absolute minimum requirement for new product development success. Suggested elements needed for a good product definition are given in the NPDP matrix in the Appendix. Where possible, inputs should be solicited from key players across the NPDP matrix (documentation, product support, sales, product marketing, quality assurance, manufacturing, finance, system engineering, and software engineering and hardware engineering) that are likely to be responsible for the planning and execution of portions of the project. The early involvement of key players will help ensure later success. The project/program manager is responsible for the product's definition and all the steps involved in it: synthesis, collection of data, organization, presentation, communication, and obtaining final acceptance for it. These individuals must possess certain characteristics that enable them to work proactively within their organization to develop the feel necessary to lead and manage the undertaking.

A product definition, represented by a definition document (external reference specification), measured prototype performance, and a draft plan produced by the horizontal organization, is evidence of product definition stability. Some change is likely to occur, but the product development team is now in a better position to assess the impact of making any change to the definition as it develops. With a documented definition in place, the process begins to exhibit certain means by which it can be controlled.

"Creeping" Features. It is not likely that "creeping features" will ever be stopped. Too many forces are at work to ensure their existence, but they can be brought under control. At the intersection of the market information row and the product column of the definition matrix, the request is made of the product development team to agree on *necessary* performance, features, cost, price, and timing. A change in product feature set is likely to influence the performance, cost, price, and timing of the product. Change to the original definition should be executed only with close coupling and interaction between functional members of the team. Research and development, marketing, and manufacturing must discuss the desired change, why it is necessary, and what is possible within a given time reference. Compromise is

needed by all participants. Good business judgement must dominate. At times, certain team members will have to be very flexible to make necessary changes a reality. At other times, the program will progress without alteration to the product's feature set. The team has to make difficult choices followed by rigorous work.

Using the external reference specification as a control document makes the appearance of feature creep highly visible throughout the new product development environment. Any change, regardless of size, should be made apparent to all matrix team members so they can assess the impact it will produce on their work output. The program manager then collects team feedback to determine the potential impact of the change. The costs of making the change can be compared to the returns that will be produced in the marketplace. Based on the cost-benefit analysis, a decision is made. If change is viewed as desirable, the ERS is revised.

A cost-benefit analysis is easiest to perform on me-too-with-a-twist and next-generation product development efforts. For these classifications, a fair amount of market knowledge is available, making it simpler to assess the sales gains produced by the added feature. Market research techniques (sales research, structured surveys, focus groups, interviews, etc.) can be put to good use to include the customer's desires in the decision-making process.

Cost-benefit analysis is difficult to do on first-of-a-kind and derivative development efforts headed for new markets. These efforts are also the ones most subject to the creative instincts of those involved in the birth of the new product and market concepts. Having completed a definition matrix for these types of efforts is a major step toward keeping "creep" change visible and controllable. Visionary people now have a product and process model, existing outside of their mind, that they must physically modify. This forces them to think through the cause-and-effect relationships of their actions. If they are prudent, they will do the headwork before suggesting a change in the feature set of the product. Making feature change on first-of-a-kind efforts should not be made extremely difficult to do, since it is sometimes necessary, but a documented product definition forces those advocating change to face the full implications of their actions.

Ineffective Decision Making. When viewing the definition matrix, it becomes easy to visualize the magnitude of the decision-making problem. Just how many decisions are involved in creating a good product definition? Viewed from only an analytic perspective, the minimum number of decisions required for even the smallest of efforts becomes very large. Therefore, it is necessary to structure the process to facilitate easy decision making (i.e., decisions involving little or no conscious effort) and to concentrate precious time and resources on the key issues. The definition matrix becomes the guide to making quality new product development decisions by making those items that could be significant routine and by making those items requiring the greatest attention highly visible.

To use the definition matrix concept effectively, new product developers need to have "real time," accurate, and appropriately filtered information. Such information can help in spotting problems and opportunities early on. It will help a number of participants within the organization develop the intuitive grasp of the business that is necessary to be responsive, effective, and, most of the time, right! This information can also be used, by those who prefer the analytic style, when the situation encourages a more thorough approach.

While processing real-time information, decision makers are encouraged to build multiple, simultaneous alternatives. This is done to inspire making decisions from a comparative perspective. Doing so encourages the production of quick, relative analysis; makes alternatives stronger by incorporating the best of the other ones; and bolsters confidence that the best alternative has been selected. Furthermore, decisions made on a comparative basis inherently have fallback positions that are developed as a result of the process. As you have already seen, decisions made in just about every cell of the definition matrix can benefit from the use of this technique.

To use the intuitive style of decision making, a style that is used when information quantity is low, information quality must be high. Prompt, accurate decisions usually require the utilization of a consensual process led by an intuitive-style decision maker. The consensual environment blends together odds, intuitive, and analytic decision styles—a mixture that can view issues from varying perspectives. While certain individuals will approach making the needed decisions from the odds-based and analytic perspectives, those with the intuitive style must lead time-sensitive consensual processing. The odds-based individuals tend to be too fast; the analytic people take too long. Only the intuitive decision maker can be the effective integrator. This individual collects input from all participants and engages in meaningful interaction to distill relevant information from the mass of data. In support of the group, or possibly as an independent action, the intuitive decision maker may seek the advice of an experienced counselor. This action builds confidence in the choice of an alternative while providing another forum to explore ideas and options. Results from the counseling session can be merged with those of previous group sessions as the final decision nears fruition. With all information present, the team makes its consensual decision. If it cannot, the process of "consensus with qualification" is used, allowing the lead decision maker the opportunity to make the choice, with participants understanding who made the choice and why.

Making effective decisions using a well-functioning process in the definition phase enables use of the same techniques in the future. If poor decisions are made within this phase of the process, it is not likely for the situation to improve as the new product development process moves to its next phase.

Technology Changes. The technical people involved in setting the product's definition have a difficult balancing act to perform. They must select a technological sword that gives a competitive advantage without excessive cost, as well as one without excessive developmental risk. In addition, they must look ahead in time to make certain that the technology that was selected is likely to remain competitive when the product reaches the market. These factors need to be addressed and incorporated in the definition of the product.

For me-too-with-a-twist products, technology selections must be made almost entirely on time-to-market considerations. Selections are made up front and they must be right! Product development must stay in phase with the cycle time of technology. If Intel and Motorola produce a new generation of microprocessor every 18 months, then it is necessary to have products based on that technology follow suit within a few months of supplier introduction to remain in the competitive race. It's becoming well-accepted knowledge within the cyclic product development environment that a few months' savings on each product iteration can produce a 6- to 12-month advantage by the time the third turn takes place. This time advantage is generally sufficient to gain a dominant competitive position. The definition matrix provides developers of me-too-with-a-twist products with adequate checks and balances to select the appropriate technology up front so that it will not require revision in the planning and execution phases—this type of development cannot tolerate a technology change and still remain competitive.

Other efforts with reduced time sensitivity are likely to assume more technological risk. It is not unusual to have technological development and product development occurring simultaneously on first-of-a-kind, derivative, and next-generation product efforts. The definition matrix provides the initial structure to assist in the management of this potentially volatile situation. Later, elements will be added to assist in the planning and execution phases of the project. A product development team that completes the new product development process matrix will have identified most technological risk areas and will have thought through most of the management details associated with moving forward aggressively.

Patience Runs Out. This is usually the result of too much change too often. People working on the project get frustrated watching their work get canned and redirected. Some of this will be tolerated, but extremes result in the loss of key personnel which, certainly, will cause further disruption and delay to the project. A good definition keeps the organization happy and motivated—the definition matrix helps ensure that a quality product definition exists for the entire horizontal and vertical product development team.

Project/program is Too Big. People who tend to be good at defining the work that a project team must perform also tend to be aggressive and optimistic. Those who choose to move forward on a project without a stable definition will sometimes discover, when it is too late, that they bit off more than they could chew. It is at this point of discovery that the project will either be cancelled or, most likely, grow in an uncontrolled fashion. Previous errors in definition tend to create new errors in definition. Once a project/program enters this trap, there is usually no graceful way out. It becomes very visible within the organization and too many people try to get involved in helping out. The best action is to avoid the situation from the beginning. The definition matrix provides the structure to instill the necessary discipline to avoid making this mistake.

The estimation technique that links time to market, productivity, work complexity, and staff size provides a convenient method to make reasonably accurate estimates on program size. Any individual estimate may have significant opportunity for error, but the technique works on collecting data from throughout the organization and averaging a collective group of estimates. It requires that horizontal and vertical inputs contribute to the estimation process. Surely, a structured estimation process brings those making predictions into closer contact with reality. Keep in mind that these estimates will be fed forward to the planning phase, where detailed planning will produce a new set of time-to-market results. The estimates from the definition phase will be used as a comparative point of reference to measure the output from the planning process. Adjustment will be made to arrive at realistic targets for the execution phase.

Changes for Manufacturability. Somewhere near the end of the project, a key member of the manufacturing team may discover that not all elements of the design are readily manufacturable. Of course, this kind of problem at this time in the development of the new product is something that most on the team would have preferred to avoid. It comes as a surprise to everyone in design and development that they could have possibly made a mistake. Information to prevent this situation was available before the commitment of resources to design the product. A well-communicated manufacturing strategy would have prevented its occurrence. Simulation and scaled prototyping may have also eliminated such a problem.

Manufacturing process models can be developed to communicate design requirements to the design team. Ideally, the manufacturing team has sufficient self-motivation to prepare the process descriptions in suitable detail. Computer-integrated manufacturing (CIM) networks can be structured to make the process descriptions very visible. In cases where CIM is not available, simple process descriptions that inform designers of process information needs, labor hourly rates, overhead rates, cycle times, and volume dependencies go a long way to helping control this variable.

A fully developed new product development process matrix serves to identify the interactions that must take place between design and manufacturing personnel. Without close coupling between the two groups, concurrent design and process development would be impossible. Even less complex items, such as integrating new designs into existing manufacturing processes, are made easier by having the two teams work closely. The need for cooperation is omnipresent in the new product development process. Interactions can be initially modeled using the NPDP matrix. Future work and information flow will be modeled using a network once the program completes its detailed plan.

Shortage of Engineers. Good engineering talent is always in short supply. For this reason, it is most important to define project work in such a form that it can be accomplished by humans who, on occasion, take time off from work to eat and sleep.

Estimates made during the definition phase of the project can be used to predict staff requirements reasonably well. An organization that has a historical software productivity of 20 lines per day had better staff a 10,000-line project with five engineers to get it done in six months. This staffing estimate is made by linking work complexity, work group productivity, staff size, and time to market as described in Section 3.1.6, "Product Definition Output." Once staff estimates are known, it becomes possible to determine if a shortage exists or when engineering resources will be available to work on the new project.

While exact assignments may not be known in the definition phase, work breakdown structures do begin to materialize. Top-down estimates of staff quantity lead to preliminary thoughts on work distribution. Managers will start thinking about the magnitude and timing of work to develop some feel for the accuracy of the estimate. If the estimation process predicts that five engineers will be needed, questions are naturally raised on the work that the five engineers will perform. A bottom-up thought process begins to surface. It's usually at this point that identification and recruitment of talent become major issues. Bottom-up thinking suggests that personnel is likely to be available to do jobs A, B, and C, but not to do jobs X, Y, and Z. This can lead to change in work definition or, most likely, to the beginning of the recruitment process to fill openings identified in the preliminary breakdown of work.

Most engineering guesses are highly optimistic and result in engineering teams taking on more projects and work than they can handle. Developing good engineering estimation techniques and skills is the best defense against taking on too much work for too few people. Good estimation techniques help get the number of people right and set realistic work-load and time-to-market expectations for the project team. Following the structure provided by the definition matrix ensures the production of quality estimates by the new product development team.

In the planning phase, managers will deal with balancing engineering assignments over time, identifying assignments where new recruits will be placed, and distributing engineers over multiple projects. For now, the new product development team has done its job in the definition phase.

Shortage of Resources. Thinking about product development from just a systems engineering, hardware engineering, or software engineering perspective is too narrow a focus and yet many organizations expect the responsibility of developing new products to lie solely with the research and development manager. However, as can be seen from the definition matrix and the new product development process matrix, horizontal and vertical interactions are necessary to make new product efforts successful. R&D, marketing, and manufacturing support groups need to be fully included in the overall functionality of the process. Each matrix cell implies the existence of a support network of people that must be present to produce new products.

Too much work with too little support is a familiar cry. Rarely will one project operate in a vacuum, away from all others. The definition of one project can have a big effect on the entire organization if it is allowed to consume all support resources. On the other hand, it is possible to create definitions that are realistic for both the project and the organization. From a quality definition, the plans and projections of future work that support people need to perform will be synthesized. The sum of the requirements from all projects can be compared with available capacities in order to make the appropriate adjustments. Feedback from support groups can be used to better manage the flow and timing of work. In general, most support groups can be highly productive if they are given some advance notice to prepare for work. The definition and NPDP matrix models provide support groups with early insight into the operation of the new product development process. Surprise is eliminated. Process visibility is provided and team interactions are defined and encouraged.

Shortage of Design Tools. In the development of high-technology products, design tools are almost always required. Individuals working without the latest tool technology often feel they are working at an extreme disadvantage and will speak up about their situation. Tools are appearing at a frenzied pace from suppliers around the world. Hardly a design team works without desktop publishing systems, mechanical design tools, electrical circuit simulators, materials' databases, software development systems, integrated circuit (IC), and personal computer (PC) design systems, and other tools too numerous to mention. Computer-aided design/computer-aided engineering (CAD/CAE) systems are having a major impact on the way new product development efforts are being performed; they have the potential to streamline the entire new product development process. Given such circumstances, product developers are constantly adding tools to the process to make improvements in productivity possible. Frequently, the focus on productivity in isolated areas does not really help the overall development of new products—I learned this lesson the hard way in the mid-1980s.

At that time, a new tool, the silicon compiler, appeared on the market. A few members of my work group saw the potential benefits offered by such a tool. In certain applications, we could take entire logic boards and reduce them to a few chips. Parts count would decline and manufacturing costs would be dramatically reduced. From a development perspective, the cost to design, fabricate, and test a chip was about twice the cost of following the corresponding process in conventional printed-circuit technology. The end product could be manufactured at a fraction of the cost of the PC board. We convinced ourselves and our management team that the silicon compiler would be a welcome addition to our stable of design systems.

All the claims the tool manufacturer made were true; the silicon compiler we purchased worked flawlessly. This fine tool, however, had no impact on our output of new products. Why, you might ask? First, it did nothing to reduce time to market. As it turned out, my work group was time-limited by definition issues and not by execution issues; naturally, the tool was unable to help us decide what to do. Second, as the definition of our target product stabilized, we learned that an alternative architecture was required that could not be produced on the silicon compiler. Third, others that had initially expressed interest in using the tool continued to work with tools familiar to them rather than utilize something new. We had jumped the gun in bringing this tool on line and had gotten burned!

The product definition must accommodate tools that are available to the project/ program, as well as those that need to be acquired. Available tools, especially the very expensive ones, frequently need to be managed as scarce resources. Tool acquisition will be based on utilization and the economic performance of the asset. It will be necessary to justify the true cost of the capital investment. The definition matrix and new product development process matrix help make the financial justification a simple mathematical exercise. Payback period, average return on investment, or discounted cash flow can be computed for each tool alternative as the tool's impact on project time to market is assessed.

Working through the matrices makes time to market a highly visible parameter. Generally speaking, tools that shave months off the new product development process time are very easy to justify. Those that produce improvements of only a few weeks can also be found to have solid financial benefit. Individuals who feel that design systems are in short supply can be asked to perform the financial justification based on the *overall* TTM impact.

Another benefit provided by the matrix structure is the visibility it gives to the *overall* new product development process across the organization. This visibility can often save great sums of money, as in cases where tools might have been added to speed up a part of the process that will have no impact on the overall time to market. Making process paths that are not critical to work faster is often unnecessary and certainly not a priority. Obviously, critical path times need to be reduced first. As reductions are made, other critical paths that require attention will emerge. The definition and NPDP matrix help segregate process zones where tools will help reduce time to market from those where the tools will not have an impact.

The phenomenon that is described above, while only somewhat visible in the definition phase, becomes clear in the planning phase. Tools that can be shown to impact the critical path and save months of time can be acquired during the definition phase. Acquisition of those tools whose impact is not immediately obvious should wait until a detailed project network emerges and the planning phase is nearly complete.

Lack of Management Support. It is hard to support something that is not clearly understood. At the same time, it is difficult for the project team to develop links with managers that can improve communications and thereby gain the support that is needed. The definition matrix and resulting output from the new product development process matrix generate support for the new product as it evolves. People are linked both vertically and horizontally within the organization. Management is primarily tasked with the production of the business definition. Depending on organization size, this definition can represent a single entity or it can be the collective sum of many divisionalized operations. A single business definition exists for the entire corporate structure. Quite likely, each division within the corporation will produce its own. Business definitions make it much easier to realize product definitions.

People supplied with business definitions and given the job of creating product definitions have already received management support. Product developers are not working in a vacuum. Business leaders are not disconnected from the creation of new products. All members of the organization are working together to make new products appear in the market.

In many cases, communicating to management with prototypes of the end product is an effective technique. It is easier for people to develop an understanding of things they can see, hear, touch, and feel. For this reason, the output items suggested within the NPDP matrix meet the see, hear, touch, and feel criteria. Creative use of communication media (written, audio, video, etc.) is also recommended to enhance management and project communication. Again, the first step in gaining management support is a good definition: one that is well communicated and well understood.

Miscommunication with the Customer. Provision is made in the definition matrix to gather and act on customer feedback intelligently. A number of checks and balances are in place to keep individuals in various functional areas informed of customer desires. Assuming all functions represented horizontally on the new product development process matrix have the potential to make the product better for the customer gives added meaning to the matrix model. It is now possible for customer information to enter the organization on any one of 12 points and have the new product development team act on the information. While input can be received by each function, the NPDP matrix makes it clear that closure with the customer occurs only through the program management cell after the matrix team has had the opportunity to review the revised external reference specification. A system that accepts multiple inputs while providing only one output to the customer makes miscommunication quite difficult.

In addition to channeling the flow of information, product classification concepts help the organization better understand how to collect information from customers effectively. Figure 3.2 provides suggestions on how to involve customers with products that are being "pulled" to the market. The guidelines suggest that actions taken for a "pull" product are very different from the recommended actions needed for a "push" new product development effort. Getting all product- and customer-related interactions understood and taking the needed actions to build the correct communication channels with customers then becomes a team exercise that is guided by relevant aspects of the definition matrix and the NPDP matrix.

Internal Miscommunication. Small project teams have trouble with internal communication. Large programs have even larger problems in this area. Over the course of a new product program, it is not unusual to deal with thousands of tasks. Each of the tasks will have predecessor and successor relationships. There is so much information to communicate! The new product development product matrix is structured to solve the miscommunication problem in the definition phase.

Key responsibilities of work groups are identified. The tendency to say "it's not my job" has been eliminated. Each functional group knows what it is supposed to do. Prototypes and simulations of the final output have been produced to cement the vision of the product in the team's mind and provide a bond that can carry the project from beginning to end. Workers have even gone on to define work complexity, determine work-group productivity, estimate staff size, and compute time to market for their portion of the product development program. Involvement of this type makes it almost impossible for miscommunication to occur.

As the program moves forward, network modeling techniques will be added to make continued improvement in the quality of information flow between functional teams.

Difficulty Communicating Concepts. This delay production ingredient is almost always associated with first-of-a-kind product development efforts. Emphasis on the external reference specification helps the development team communicate the product concept. Writing the ERS is a good exercise to help focus thoughts and stabilize the definition. Creating the ERS has the potential of being a frustrating experience. If it is, it is usually because there is a lack of stable information, as well as frequent change in alternative synthesis and analysis. Change and complexity dominate the initial portion of the definition phase. The soup example helps illustrate how difficult it is to tell someone what the soup's color, texture, smell, and taste are if it does not yet exist. An experienced product development manager starts with an ERS skeleton. As elements of data surface, they are added to generate a more complete structure. Frequent change is expected and anticipated. Given the proper incubation time, a complete ERS will emerge.

For the ERS to be fully effective, it must be supported by simulations, prototypes, charts, sketches, video screens, and pictures. These vehicles assist in communicating what is difficult to transfer in written form and yet very easy to transfer in visual and physical form. Only highly skilled writers can convey in writing how the soup tastes and smells so that *all* readers would interpret the writing the same. A poor writer would have difficulty communicating the concept of soup! Yet it's very easy to convey taste and smell information if soup samples are available. Skilled writers of an ERS can identify the areas that need the support of visual and physical information transfer. They will create a blend that meets the horizontal and vertical information needs of the organization. The new product development process matrix identifies the need for an integrated presentation of the product's definition.

Money Runs Out/Project Cancelled. Most of the time, the reason projects get terminated is because the global definition missed a key point or two necessary for the successful completion of the effort. Of course, it is best to cancel a new product undertaking during the definition phase than at a later time. Likewise, the definition phase is the time to place concerted emphasis on developing a thorough understanding of product and process. This falls on the theme of working on the "right things." If a project/program is "right" by definition and receives strong organizational support, it can generally withstand hard blows in the planning and execution phases and still come out a winner. Cancellation is not likely for a project that gets go-ahead approval at the completion of the definition phase—many things had to have been done right to get this far!

Also, some projects are cancelled because they were never effectively started. In organizations where it's more expedient to run a large number of new product development efforts well below their true full staff needs, all efforts run the risk of being late to market and are fully exposed to the associated risks. The definition matrix attempts to guide new product development teams in the direction of understanding full resource needs to develop the product in the requisite time and having the necessary attributes. Each cell of the matrix encourages the creation of information that is used by the project team and management team. Vertical and horizontal linkages are built as the entire organization becomes committed to taking the product to its logical next step. As detailed knowledge of each product development effort builds, developers and management begin to see the means by which some will be accelerated and fully staffed while others are held back. Making such decisions now prevents the organization's investment of time, money, and human resources in an unworthy project. Remember, new product development efforts that exit the definition phase should not have consumed any more than 20 percent of their development funds.

Product Looking for a Market. By definition, all first-of-a-kind products are looking for a market. Some product developers have a highly focused and accurate view of the market. Others may not, and the product may arrive on the market with no one ready to buy it. This usually happens when a team of technical people develops a product that offers excellent performance but is targeted at an esoteric market segment. The team's attempts to "push" the product on the market may not be working. To avoid the situation, developers of first-of-a-kind products are encouraged to support their visionary efforts with close one-on-one interaction with potential customers and to consider the utilization of focus group techniques to try product concept on customers while in the definition phase. Doing the up-front work outlined in the market information cell of the definition matrix can help the team avoid producing a major marketing mistake.

If, however, a team finds itself in this uncomfortable position, a recommended next step would be to make an attempt to complete the definition matrix after the product has been introduced. This activity can sometimes lead to a very pleasant turn in business. The exercise will identify reasons for poor market acceptance of the product. With this knowledge available and with some luck, the team may be able to make a few adjustments to the product and its processes, sometimes turning a disaster into a success. Some front-end discipline even late in the process can work to identify large potential market opportunities for the product. In other cases, an examination of the best effort product definition will lead to a decision not to pursue the opportunity.

4

PLAN SYNTHESIS

In this chapter, you will learn about

A product development plan synthesis technique.

An easy and effective means to allocate resources to projects.

A visual, fun method for doing cross-functional planning with simple tools.

The creation of the new product development process (NPDP) network model.

Assignment of resources to tasks.

Estimation of task work quantity.

Synthesis of the project task network.

It's easy to get players. Gettin' 'em to play together, that's the hard part.

Casey Stengel

4.1 PLANNING PROCESS

A brief but intense planning phase is initiated when the new product's definition begins to achieve some degree of stability. As Figure 1.1 shows, the planning phase lies between the definition phase and the execution phase. It is the bridge between new product definition ("right things") and new product execution ("things right"). It is a unique phase within the new product development process. Only when the new product development team thinks the definition is complete can they begin the production of a serious plan. Similarly, in a cross-functional, concurrent engineering environment, the execution phase cannot begin until a plan, fully supported by the necessary resources, is in place. The planning phase should take somewhere between two and six weeks. Effective use of this time sets the stage for the production of the new product within the execution phase.

This chapter offers a plan synthesis alternative that makes the process highly visible and improves the overall quality of the resulting plan. The output from the plan synthesis process drives plan analysis, which is discussed in detail in Chapter 5. Following the process and suggestions in both chapters will speed and simplify the planning process. The links between definition and execution will materialize, and the team doing the planning will actually have some fun with the process.

4.1.1 Overview

Figure 4.1 describes the flow of information through a typical planning cycle. The entire process must be activated each time a new product development effort enters the planning phase. Initially, two paths are followed. The first relates to resource allocation and the second to the creation of a hierarchical network model of the new product's execution phase. Resources need to be allocated from the organization's global pool to projects; at the project level, they will be assigned to tasks. Once applied at the project level, knowledge of resource utilization improves, resulting in the refined allocation of resources between multiple projects. Resources have a fixed and variable cost associated with their utilization that can follow them into a data table. Likewise, it is relatively easy to link resource availability to a calendar to establish each resource's accessible time. The modeling path leads to the development of a complete network model of the project. The modeling process begins with the construction of a hierarchical representation of the project and terminates with

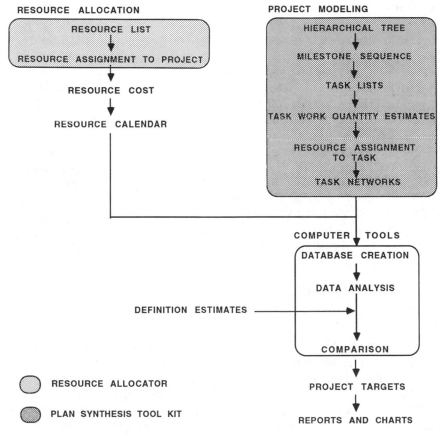

Figure 4.1 Planning process flow diagram.

the generation of a task network diagram for each branch in the hierarchical tree. A project's hierarchical tree and the function axis of the new product development process (NPDP) matrix are closely related. Their coupling facilitates the production of combination matrix and network models. Information from the resource allocation and modeling paths is combined within a computer system to create and then analyze the project's database. Feedback from analysis is used to make adjustments in both paths until satisfactory targets are produced. Where necessary, changes may have to be made within other projects depending on the same resource pool. Also note the utilization of results from the definition phase in the plan analysis section.

Two new tools that I've developed for plan synthesis are presented in this chapter. A number of new plan analysis methods are presented in the next chapter. One synthesis tool is referred to as the resource allocator (RA) and the other is known as the plan synthesis tool kit (PLASTK). RA was designed to expedite and simplify resource listing and resource allocation to projects. PLASTK greatly simplifies the steps needed to convert task listings into task networks. The contributions of these tools to the overall process are shown in the shaded areas of the planning process flow diagram. RA and PLASTK outputs are structured in such a way as to make entry of data into computer systems a relatively straightforward task. Another interesting aspect about each of these tools is that you make them yourself. A third tool, plan analysis software (PAS), runs on industry standard personal computers and is designed to be used with the popular brands of planning software on the market. PAS analyzes the project's database using a number of analytical, comparative, and graphical techniques to assist planners in the evaluation of the database. PAS is described in detail in the next chapter. All the aforementioned tools were crated to ensure the production of quality results from the planning process.

The fundamental ingredient in the production of a quality plan is knowledge of the desired outcome; planning is done to establish a method for the cross-functional realization of the new product. Knowledge of the desired result was produced during the definition phase along with many of the detailed requirements of what must be done during the execution phase. The planning process determines the structure, resources, timing, and cost of realizing definition-phase objectives.

Attention now focuses on the resource allocation portion of Figure 4.1.

4.2 RESOURCE ALLOCATION

4.2.1 Resource List

The resource list is a complete compilation of all engineers and support professionals that work on the development of new products. It includes development, marketing, manufacturing, sales, service, and support engineers and their managers, as well as the professionals from personnel, finance, and quality. The list does not include all of the individuals that are in place to support those on the resource list; those not on the list, for instance, include top-level managers, functional-level managers, hourly manufacturing personnel, clerical workers, and so on. The goal of the resource list is to identify the top 20 percent to 30 percent of the new product development work force that keeps the remaining 80 percent to 70 percent gainfully employed. This distinction is important in keeping the number of individuals on the list at a reasonable level.

For example, an operation with 1000 employees who contribute to the development of new products should have about 250 people on the resource list. The breakdown might consist of 100 research and development engineers, 5 project/program managers, 70 manufacturing engineers, 50 marketing and sales engineers, and 20 professional support personnel. Optimal deployment of these resources is a critical determinant of new product development process success. The assignment of 250 well-paid professionals is not a task to be taken lightly!

Where possible, the list should be sorted by function, job classification, and name. New hires and individuals to be hired should appear on the list at the functional level. The number of new hires should be in close alignment with the projected human resource growth forecasts of the organization. At this point in the exercise, it is not necessary to know the job classification of each potential new employee; we'll let the assignment of resources to project and task help determine the skills that need to be added to the organization. In this way, the planning phase will become a key tool utilized in identifying the resource skills needed by the organization.

Generating the resource list is an awareness-expanding exercise for just about everyone involved with the planning process. For the first time, an organization gets a complete horizontal and vertical view of all the people that hold critical positions in the development of new products. Individuals on the list and managers looking at the list begin to appreciate the number of decisions necessary to make the assignments effectively; it becomes clear that only collective brainwork, with a bit of tool support, can attempt to solve the allocation problem.

4.2.2   Resource Assignment to Project

One of the key reasons for doing the product definition makes itself evident at this point: the resource allocation people have a relatively clear view of the resources needed by the project. Information about the resource needs of the project is utilized to generate the names of the individuals, from the resource list, who will be assigned to work on the project should it successfully exit the planning phase. In addition to these names, for planning purposes the project planning team is given the likely time that these people and other resources will be available to commence working on their new assignment. Individuals doing the allocation will not know specific assignments at the project level, but they will know specific skill types (electrical engineer, software engineer, mechanical engineer, etc.) and the numbers of personnel needed by the project.

Information leading to the production of these critical resource allocation decisions began to surface at the output of the business definition process. Throughout the definition phase, key decisions were being made by the definition team about future resource needs. Informally, the management team responsible for allocation of resources to projects has observed the definition phase and provided some ongoing guidance on resources likely to be available as knowledge of people, processes, and product advanced. Now, with a full working knowledge of resources needed by the project, it is in the best position to observe how other projects, well into their definition or execution phases, are progressing to develop the means by which resources will transfer to the new product effort being planned. Thoughts on how to get the project fully staffed begin to stabilize. This gives resource allocation specialists a clear idea of the desired sequence of new product introductions scheduled by the organization and some insight on resource allocation alternatives that can support the overall new product process.

Evidence of this process within the MONI example can be found by tracing the product selection process from the product-line guidance section of the corporate business definition to the necessary products and services section of the division business definition. Refer to the Appendix to review the corporate and division business definition examples. Also note that estimates of resource requirements were identified and listed on the new product definition sheet for each product development effort. Key information (which projects to consider, staff requirements per project, etc.) to assist in the allocation process is available to the allocation team.

To initiate the allocation process, each function (research and development, marketing, manufacturing, etc.) assembles a resource allocation team and places the resource allocator (RA) tool into action. To make your own RA, all you need are D-size sheets of paper and some small Post-its or other such tags or "tokens." A token is assigned to each resource for each allocated time period. Resources are then assigned to projects across the page. This is a closed system. All resource tokens must be utilized and no two identical tokens are allowed to occupy the same column. For instance, if allocations are being done on a quarterly basis with a three-year rolling window, each resource is represented by 12 tokens. The D-size paper is marked into 12 time bins along the horizontal. Projects are identified on the vertical. Twelve tokens for each resource are assigned to selected projects as they are placed horizontally across the page.

Figure 4.2 illustrates the structure and utilization of RA. In this example, three years of time is segmented by quarter, so the resolution of resource assignment is in three-month intervals. The resource pool over this time period is formed by representing each resource by a token within each quarter. Thus, five resources are available in Q1 through Q7. In Q8, a new person is added bringing the total count to six. The resulting resource pool has 65 tokens representing available resources over this period. Next, resource allocators move tokens from the resource pool as they are assigned to projects. The figure shows the pool is used to keep two projects in operation. Project A is staffed with four individuals while the fifth works on the definition of Project B. As Project A nears completion in Q7, resources shift to Project B. Q7 and Q8 become a period of transition as resources leave Project A and begin contributing to Project B. Additionally, a "new" person is hired in Q8 to work on Project B. The RA concept is very simple. This inherent simplicity makes it an extremely functional and versatile tool.

Meaningful resource allocation decisions can be produced only when decision makers have the opportunity to interact and discuss the placement of each resource

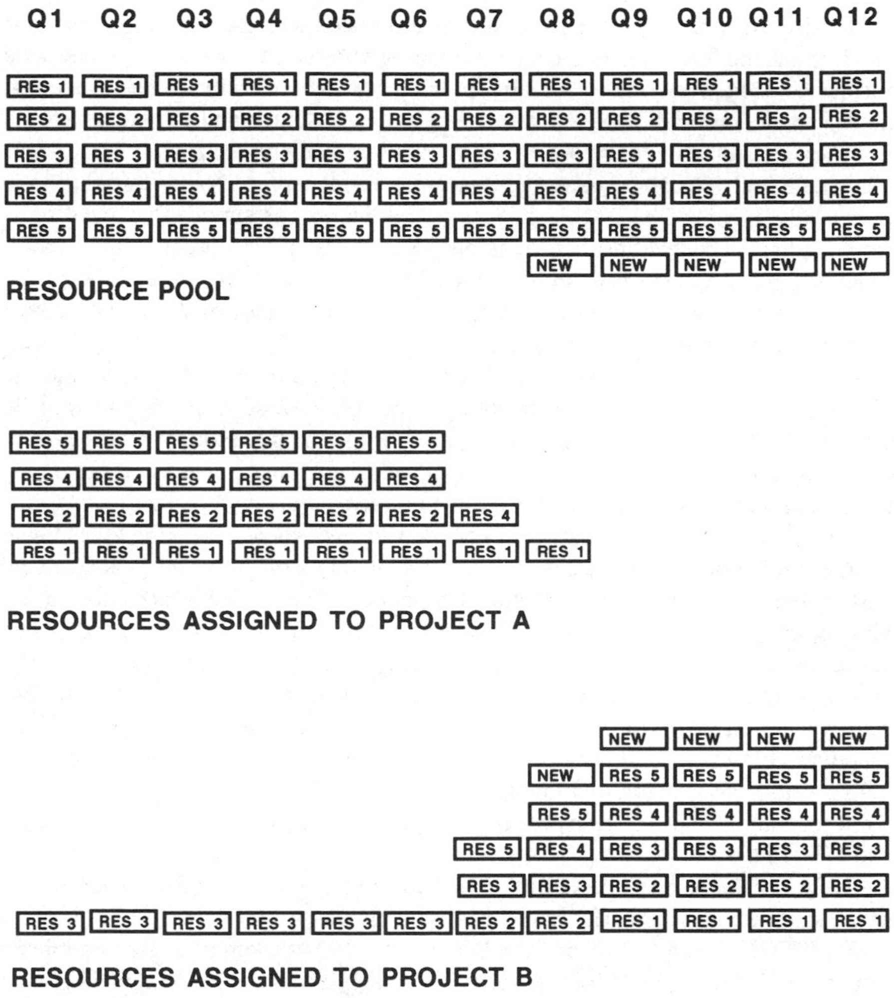

Q1	Q2	Q3	Q4	Q5	Q6	Q7	Q8	Q9	Q10	Q11	Q12

RES 1	RES 1	RES 1	RES 1	RES 1	RES 1	RES 1	RES 1	RES 1	RES 1	RES 1	RES 1
RES 2	RES 2	RES 2	RES 2	RES 2	RES 2	RES 2	RES 2	RES 2	RES 2	RES 2	RES 2
RES 3	RES 3	RES 3	RES 3	RES 3	RES 3	RES 3	RES 3	RES 3	RES 3	RES 3	RES 3
RES 4	RES 4	RES 4	RES 4	RES 4	RES 4	RES 4	RES 4	RES 4	RES 4	RES 4	RES 4
RES 5	RES 5	RES 5	RES 5	RES 5	RES 5	RES 5	RES 5	RES 5	RES 5	RES 5	RES 5
							NEW	NEW	NEW	NEW	NEW

RESOURCE POOL

RES 5	RES 5	RES 5	RES 5	RES 5	RES 5
RES 4	RES 4	RES 4	RES 4	RES 4	RES 4
RES 2	RES 2	RES 2	RES 2	RES 2	RES 2	RES 4
RES 1	RES 1	RES 1	RES 1	RES 1	RES 1	RES 1	RES 1

RESOURCES ASSIGNED TO PROJECT A

								NEW	NEW	NEW	NEW
							NEW	RES 5	RES 5	RES 5	RES 5
							RES 5	RES 4	RES 4	RES 4	RES 4
							RES 5	RES 4	RES 3	RES 3	RES 3	RES 3
							RES 3	RES 3	RES 2	RES 2	RES 2	RES 2
RES 3	RES 3	RES 3	RES 3	RES 3	RES 3	RES 2	RES 2	RES 1	RES 1	RES 1	RES 1

RESOURCES ASSIGNED TO PROJECT B

Figure 4.2   Resource allocator (RA) example.

while keeping the view of all remain projects in the foreground. In a typical organization, RA is first used by the top management team of the R&D function. The R&D functional manager and a close team of associates make an initial attempt at allocating their function's resources between the various projects in operation within their department over the planning time frame. Members of strategic business units who work within R&D are actively involved in the allocation of resources between projects within their function. These individuals then have intimate knowledge of how projects can flow through R&D over time. They then take the R&D staffing profiles to their respective SBUs to initiate the resource allocation process in marketing and manufacturing.

The marketing and manufacturing SBU representatives, after discussion with the

R&D SBU member, take the R&D information back to their respective functions and begin using RA to distribute the resources within their work groups between the various projects over time. Naturally, problems will be identified. The functional management team will have to interact and resolve resource and project issues. As it makes decisions, it will be necessary for it to interact with the SBUs. Discussions take place and allocation issues get resolved. Evidence of problem resolution is graphically displayed by RA. Arriving at a solution may be simpler than the discussion implies. The net result is that the tokens get moved around by a cross-functional management team until participants are satisfied with the placement. The people doing the allocating and moving of these tags consist of the organization's top management team and managers of its SBUs.

Frustration with the use of a number of manual and computer tools to allocate resources to projects led to the creation of the RA system. Working individually with local information and sheets of paper was fine for the allocation of resources within my direct control; however, the paper technique broke down when attempting to use it to allocate resources within a function, across functions, or across projects. Computer tools were tried, and again little was gained when attempting to deal with a large number of allocatable elements and a relatively large number of management participants. The screens were too small to display the needed information in full view of the decision makers. Data input parameters were too numerous and lacked connectivity between one data file and the next; nothing was in the system to make allocation over a given time an error-free process. Calculation time became significant when attempting to balance and level resources across multiple projects and functions. I found myself becoming a servant to the tool rather than having the tool help me allocate resources and merge my output with others performing the same work for their group. A better way to manage the allocation of hundreds of people had to be found.

RA provides the basic capability to make changes instantly and view the results at the same time. This performance is absolutely essential for full utilization of a team of talented people on a project. People using this system need about a minute or two to fully understand how it works. Once they do, the token system begins to be widely used, and the ensuing discussion and interaction invariably lead to the generation of the final output.

It was an absolute requirement that the tool provide full visibility of where resources were assigned within a function. The functional managers of R&D, marketing, and manufacturing should know where each member of their respective staff is assigned. Ideally, this information should be visible in graphic form, so it can clearly indicate present staff allocation deficiencies and help identify where future problems might occur. RA gives the functional allocation team this visibility through the use of D-size sheets of paper hung on the wall. There is no limit placed on work space by the size of the CRT. The work space grows in proportion to the number of resources being allocated. Cause-and-effect relationships across an entire functional area become obvious. Managers view how each project is utilizing resources within their function in graphical form; a resource histogram is created for

each project as resource tokens are allocated to the projects. The formation of resource histograms can be seen in Figure 4.2. Resources can be moved from one
project to the next simply by moving tokens. No computers are necessary; calculation time is limited only by the speed at which tokens are placed and moved.

Cause-and-effect relationships become visible as soon as allocation managers
from different functions begin the assignment of resource tokens to time bins. The
managers are forced to structure the time flow of new products to market. Projects/
programs will have time-to-market priorities established. They will address the
number of projects that can be supported concurrently. Quite often, adequate resources will not be available to cover all work items in a parallel fashion. This will
necessitate the placement of projects in tandem. Allocation managers will then have
to arrive at a workable arrangement of concurrent and tandem product support structures to realize new product development objectives for the organization. Thoughts
on network models will begin to surface. Predecessor and successor project/pro-
gram relationships will be established. Which project should lead? Which project
should follow? Are all projects presently planned fully staffed? Some order will start
to emerge regarding how resources will be allocated throughout time.

Program managers working in close cooperation with the functional allocation
teams can see the staffing plan for their program materialize as elements from each
functional area are summed to create the overall allocation plan for each program.
This step is easy to accomplish by aligning R&D, marketing, and manufacturing D-
size sheets for a given program in a single column. The program manager can see
how resources from various functions will be made available over time. It also becomes clear where resources will come from to initiate the project and where they
will go as the project nears completion. Resource cause-and-effect relations begin
to surface to force the decision makers into action. The inclusion of the functional
and program joint participation helps to fuse the allocations vertically and horizontally throughout the organization. For many organizations, RA will provide a first-
time graphical view of the horizontal and vertical distribution of resources over a
number of projects and over a period of years.

This columnar graphical view makes visible resource time alignment problems
between functions. If the resource histograms from R&D, marketing, and manufacturing are not in alignment, resource allocation work must continue until all the
cause-and-effect relations producing the timing offsets are resolved. Allocators will
move tokens in a group environment until a workable solution is realized. Tough
decisions will have to be made. Some projects will have to be delayed. New hires
might have to help get existing work done before they are assigned to a completely
new project. The people doing the resource allocation form a collective understanding of why things have to be the way they are from a complete horizontal and vertical perspective. Everyone learns how they depend on everyone else for their job's
success.

The RA technique gives instant feedback on the staff profile of a given project or
throughout a function that supports a number of projects. Visual inspection shows
the rate of resource buildup, resource sustainment level, and resource roll-off. The

distribution of resource assignments over time is immediately obvious. Peaks, valleys, and other abnormalities associated with staffing and utilization decisions become highly visible. Voids where resources are not available are rapidly identified. Lower-priority projects can be delayed to reach critical mass on higher-priority programs. Likewise, recruiting efforts can be initiated to fill upcoming staff voids. No longer is it necessary to guess at what skills are required or when they will be needed; RA puts the information picture directly in front of the people who can use it to initiate the appropriate action.

An organization that is using RA to allocate 250 resources over a three-year period on a quarterly basis needs 3000 tokens (250 resources × 12 quarters). Assuming that R&D, marketing, and manufacturing each allocate resources in equal numbers means that each function will be responsible for the placement of 1000 tokens at the project level across the organization. This may sound like a large number, but a few other parameters will help put the problem of numbers into perspective. An average person, without much thinking involved, can place Post-its or other tokens to paper at a rate of 30 tokens per minute. At this rate of production, a skilled individual could allocate 250 resources over a three-year period in under two hours. Of course, no organization has a resource allocator of this caliber, so the problem will be broken down by function and by project/program as described previously. It is not at all difficult for a functional allocation team of three or four people to initially place 1000 tokens in less than an hour. Once the initial placement is made and the visual resource pictograph is produced, it can usually be adjusted over a period of two to three hours to yield the final result. Thus, productivity with RA in a team allocation setting runs at the final placement rate of about 25 resources per hour. Adjustments made to the functional allocations once projects and programs are integrated cut the rate down to about 15 resources per hour. At this rate of productivity, horizontal and vertical assignment of 250 resources can be made in about 16 working hours. I know of no other resource allocation system that can handle a problem of this complexity in a shorter period of time.

Allocators must discuss among themselves the issues that relate to the selection and deployment of resources. Intelligent discussion is absolutely essential to make meaningful placement of resource tokens at the project level. Functional management and project/program management must effectively interact for a suitable horizontal and vertical resource allocation solution to be arrived at. The placement of resource tokens on the D-size sheets stimulates the human interaction that is necessary to make this allocation process function. RA takes a task that can be dull and mundane and converts it into a group exercise that is fun and rewarding. The tool serves to open communication paths between functions and across projects.

The assignment of resources to projects has now been performed by a team of horizontal and vertical resource allocators. RA has served to bring the organization together to make critical resource allocation decisions. To reach this point, many key hurdles were crossed to arrive at the quantity, quality, and timing of resource support each project should receive.

Maintaining the RA D-size sheets is a simple matter that can be performed in a few hours each update period. In fact, some organizations use RA only once to establish the initial allocation. Once this is done, it is possible to create spreadsheet models of where the resources have been allocated. Small noninteractive adjustments can be made to the electronic files as they occur. For organizations that elect to follow this path, it is recommended that RA be utilized at least once each year, twice if possible, to maintain the horizontal and vertical communication paths within the organization. When major changes to the allocation of resources take place (a project is cancelled, key personnel resign, a technology cannot be made to function, etc.), it becomes necessary to reactivate RA to explore all of the new horizontal and vertical implications thoroughly. Doing an analysis of this type at a critical point in time is a difficult act to perform. RA makes it possible to begin the process of recovery and place resources in positions where they will benefit the organization the most.

New, or updated, resource allocation information created using RA is made available to the project/program teams in written and electronic form.

4.2.3 Resource Cost

Data on the fixed and variable costs associated with each resource can be assembled at this time and added to the resource listing. This data will be put to use when project and resource data merge in the computer system. In a typical organization, vast amounts of accounting detail data need to be reduced to produce the single number per resource. The cost associated with each resource is derived from group averages and is intended to help estimate total project cost without having each and every cost element broken down in complete detail. For people, the cost data is expressed as dollars per hour or some other convenient expense rate. Machine costs can be estimated using the double conversion of operations per hour times dollars per operation to yield a dollar-per-hour figure for each machine.

Hourly costs per person should include the direct wage as well as all other overhead attached to the individual on an average basis. Direct hourly wages are increased by factors such as payroll taxes, benefits, travel, education, supplies, facilities, tools, and support services. Typically, direct wages are doubled to arrive at a fully loaded hourly cost per person. An individual earning a direct salary of $4000 per month ($23 per hour) has a fully loaded salary of $8000 per month ($46 per hour). Figures of this type are developed for each person on the resource list. Within a given function, costs are about the same on a per-person basis, but it is not unusual to find significant cost differences between functions.

Data from accounting systems play a key role in the determination of the figure for each resource. Since all information is based on averages, cost data must be regressively analyzed to arrive at the final figures. An effective accounting system will produce the averages for each resource once it is programmed to compute the desired information. Statistical analysis of past data points can be used to project the resource cost figures years into the future to aid in the estimation of total cost for long-duration projects.

For those working without the autogeneration of the cost figures, it is necessary to compute the figures for each resource manually. To do this, access to the following information from the organization's accounting systems is needed:

Like-group employee count   Like-group total wages
Payroll tax percentage   Benefits percentage
Facility expenses   Travel expenses
Recruiting expenses   Education expenses
Software expenses   Tool expenses
Support service expenses   Management expenses
Project-related expenses   Miscellaneous expenses

Collecting the data is no small task; building the model and doing the analysis take an even greater amount of time. Quite likely, work will be reduced to building spreadsheet models that accept the listed information above and produce the final numerical output. For this reason, it is advisable that the work in the production of the resource cost figures be performed by a cross-functional, horizontal team. As team members go through the process, they can identify the obstacles that keep the information from being autogenerated. Systematically, the obstacles can be removed to realize a fully automated resource cost-generation system. An automated system provides the added benefit of not having to go through the entire exercise each time a new estimate is required.

4.2.4   Resource Calendar

This is the most simple step in this section. All that is required is the determination of a standard work calendar for the organization. Usually, only two key pieces of information are required: nonwork days (weekends, holidays, etc.) and the hours worked per day (usually eight). Vacation schedules and detailed timing detail about the individual resources are not needed.

4.3   PROJECT MODELING

To better prepare you for reading about hierarchical trees, milestone sequences, task lists, the assignment of resources to tasks, the estimation of task work quantity, and the formation of task networks, we'll discuss the path that is followed by many individuals as their career moves from student, to professional, to entry-level manager, to a manager responsible for the performance of a large team of individuals. This discussion is intended to place many complex planning issues into a simple, easily understood, perspective.

For university students, modeling task assignments is a relatively straightforward process. Hierarchically, decisions progress from selection of university, to selection of major, to selection of annual class list, and, finally, to selection of this semester's class list. Some students make these decisions quickly; for others, each step is agonizing and difficult. Regardless of the method selected to arrive at the choices involved in registration for the present semester, the outcome is the same—classes must be attended and satisfactorily completed before the student can repeat the process next semester. Work detail does not become evident until the student attends the first series of classes and gets assignment and examination descriptions from the instructor. Once all of the information has been assembled by attending all classes, it becomes possible to complete a detailed plan for the entire semester and make difficult time allocation decisions.

A few pieces of paper, pencil, and a calendar are the tools needed to produce such a plan. The semester has two major milestones: those of semester start date and semester end date. The first is set by the start of the first class, the second by the completion of the last final examination. Between these two milestones are the activities associated with each individual class. Assignments will be due on dates established by the instructors; intermediate and final examinations are scheduled. As requirements and timing information become known for each class, it is usually sufficient to write down the output items that have to be produced on the calendar to arrive at a functional plan for the semester.

Examination of this preliminary work can be of considerable value in the establishment of daily, weekly, and monthly priorities. Periods of heavy work become visible and thought must be given to developing actions that can help smooth work loads to acceptable and achievable levels. Artificial due dates must be created to spread out the delivery of the three assignments that are due on the same day from three different instructors. Special arrangements can be made with instructors to eliminate taking four examinations on the same day. A scheduled visit back home may have to be delayed to accommodate the class-induced work load. More time can be allocated to those assignments that will consume the most time, whereas less is set aside for tasks that appear to be easy and familiar. Information gathering might take a week; the full plan will be developed in less than two hours—this is two hours of highly leveraged and productive time. The establishment of priorities makes the student face the realities of this semester's challenges. Wise use of the information greatly improves the quality of life of the student who took the time to construct a simple plan over those individuals who elect to just take the semester as it comes.

To help focus energy on a weekly basis, data can be taken from the calendar and transformed into "to do" lists for each week of the semester. Each week's list is sorted by date so that at a glance the student knows what must be done each day. A student who uses this type of system prepares 16 or so individual lists at the beginning of each semester. Adjustments are made to the appropriate list as new information becomes available. The "to do" list becomes a convenient way to measure progress: a line through an item means it's done; a list with all items lined out is finished and ready for the trash can. Measurement leads to the identification of problem areas and provides early warning to permit their correction.

Generating the first draft of the plan starts other processes that need added thought. Resources will need to be assembled for each class. A job may have to be found to cover the costs of resource requirements. Books and supplies will have to be purchased. A design laboratory class requires the purchase of components with a three- to five-week lead time. A humanities class will require that certain materials be reserved at the library to ensure their timely availability. One difficult class in the student's major may assign a task that must be performed by a team of people—the student will have to assemble the team. The process of identifying the appropriate material and people resources has started. As the process generates new data, the information is added to the calendar or the "to do" list on the appropriate date.

Elements of the process that have been described were used by many of us to complete our undergraduate and graduate education satisfactorily. We learned how to use simple tools and techniques to plan some complicated activities for ourselves. Marked-up calendars and "to do" lists became an integral part of the college experience. We brought such tools with us to our first professional assignments.

On the job, much of the college planning experience is retained. Professors and tutors are replaced by supervisors and mentors. The supervisor becomes the individual who sets the requirements of the job. Early assignments are usually highly specific and structured. Their duration is about a semester in length. Under these conditions, a new employee applies familiar calendar and "to do" planning techniques and develops a way to succeed on the initial assignments.

The reward for doing well on structured assignments is to receive assignments that are less structured and far more challenging from a team perspective. An independent contributor is in an excellent position to influence work scope and timing of his or her own work. Unfortunately, this work will be coupled to the work of others in the organization in ways that might seem complex and mystical. The individual contributor begins to think in terms of task predecessor and successor relationships, but finds building models of these interactions quite difficult when operating independently. Because of the inherent complexity in building such models, team members elect to work to a top-level plan that shows only major milestones. Of course, the time placement of the milestones was developed by the best managerial guesses possible!

Everyone on the team knows the basic plan was produced without quality supporting data, but since there is no evidence below the visible placement of milestones, no one can point to where the problem might be. Since only major milestones are being measured, time movement of the milestone must occur before taking action. After the milestone motion takes place, the team must guess on the cause and initiate the perceived needed corrections. Sometimes they get it right; sometimes they do not.

The reality of the situation is that wishful thinking has overtaken rational and objective thought processes; reality is no longer visible to the individuals caught in this mode of operation. No one wants this to happen; it's accepted because there just seems to be no other way to deal with the problem. A model is needed that couples each and every team member together through time. At an early point in my management career, I thought the development of these models was my responsibility. It was difficult making the transition from modeling and planning my own work to modeling and planning the work of others.

I struggled with the creation of project models over a period of about 10 years, trying an assortment of techniques based on a variety of project management software tools. A major advance, which had nothing to do with the tools I used, materialized in 1985. At that time I was managing Hewlett-Packard's program to develop the 8770A arbitrary waveform synthesizer. It was a development effort that was under way in a number of different geographical locations scattered across the western United States. I knew a model of the entire development effort was absolutely essential for us to achieve our development goals within this distributed, concurrent engineering, development environment. Armed with an Apple MacIntosh 512K loaded with MacProject, I set out to create the model. It took about two weeks to build an activity network that contained 200 elements. The work was done in a paperless environment, going directly onto the MacProject software. Placing the first 30 or so tasks was fun and easy. After that, the creation, placement, arrangement, and connection of the remaining 170 elements became quite a chore. As each of the 200 tasks was created, its duration had to be estimated, resources assigned, and predecessor and successor relationships established. With data entry complete, a printout was run off the ImageWriter. A few hours and rolls of Scotch tape later, a new layer of wallpaper adorned my office wall. Of course, some adjustments to the network model proved necessary on the computer, and then it was ready to be tested.

The model went before every individual who was on the resource listing for their critical analysis, comment, and revision. I explained to each person why the network looked the way it did and they explained to me what was wrong with it and what should be done about it. The entire model was revised in about a week. Where things did not look right, I called responsible people over to help me or I called them on the phone to make them a part of the decision process. Elements of a shared planning process began to appear.

This revision went out and underwent minimal changes. This baseline version of the plan was taken to functional and general management and received their full commitment of all the resources identified in the plan to execute the project. We now had a model for the entire 8770A new product development team.

For me, doing the 8770A project with a network model completely eliminated the possibility of ever doing another project without one—it's an essential element in the characterization of the new product development process. The valuable lessons learned in creating this model improved the model for my next project/program. I began to form a clear concept of how to help others model and plan their work, a function that many of us have in our careers. Personal experience must be used and expanded upon in order to direct teams of people to produce accurate, distributed plans.

A major pitfall to be avoided is that of modeling someone else's work for them. You can only develop task network models in one of two ways. The first is to let those doing the work build the model; the second is to work with those doing the work to build the model. Under no circumstances can you build the model for them! The only way to get individuals and groups to establish priorities, focus energy, understand their roles, collaborate, be creative, and develop a spirit of shared responsibility is to have them be a part of the modeling and planning process—it has to be a shared experience.

The development of a shared plan cannot be done with only a computer system. To do the planning exercise justice requires the creation and management of a great deal of information and communication between many people. Research and development, marketing, and manufacturing work in a highly interactive mode. Because of the tight coupling between individuals and the complete removal of functional partitions, the product's developers must fully understand who needs what from whom and who will do what by when. The production of the network model captures the dependencies as it describes the flow of work leading to the resultant new product. The database created from the modeling exercise becomes instrumental in the realization of time-to-market objectives. Additionally, the information that is produced from the data makes it possible to observe the process visually and graphically. You need the freedom to easily modify large network models, create new branches and networks, and see the entire picture while these changes are taking place. It's quite likely that the modeling of projects will require the visual manipulation of hundreds to thousands of elements to complete the development of a realistic hierarchical and network model.

Tools and techniques that are to be utilized should be easy for people to learn and handle. People must be encouraged to produce models that are based on familiar techniques such as writing entries on a calendar or generating a "to do" list. The team developing the shared plan does not have the time to send its members off to "planning school" to make them effective hierarchical and network thinkers. Training must be quick and accurate. Ideally, the tools and process should eliminate, or greatly reduce, the need for structured training by being compatible with the information and skills that most people presently have available. Where possible, those skilled in concept understanding and tool utilization should help develop the skills of their co-workers.

I set out to model my next program using a new technique that solved all of the problems mentioned above. The system I developed was visual, used no computers, and was extremely easy to learn—it was based on use of 3 × 5 inch cards. To initiate the process, I went back to basics. I began by emphasizing shared planning methods and techniques to the cross-functional team. Without a shared plan, the team could not ensure that its goals would be met. Without a control system, I could not ensure that the shared plan was being properly executed.

To develop the shared plan, preprinted 3 × 5 cards were distributed to each functional group (marketing, manufacturing, hardware, software, systems, etc.) and the groups were asked to complete a card for each of their remaining developmental activities. Each activity (task) was formulated in terms of a specific deliverable. The team members were asked for a description of each task, their best "bottoms-up" estimate of the task's duration, and predecessor and successor information. The 3 × 5 cards that were distributed appeared as in Figure 4.3.

Upon examination of the card, it should be clear how the system is intended to function. A card is completed for each task. The person responsible for the identifi-

```
┌─────────────────────────────────────────────┐
│  TASK:     _____  │
│  RESPONSIBILITY: _____  │
│  DURATION:   _____  │
│  NETWORK:    _____  │
│                                              │
│  PREVIOUS TASK:  _____  │
│  NEXT TASK:  _____  │
│  MILESTONE:  _____  │
│                                              │
│  COMMENTS:  _____  │
│  _____ │
│  _____ │
└─────────────────────────────────────────────┘
```

Figure 4.3   Three by five inch modeling card.

cation and creation of the task makes a preliminary resource assignment and estimates the duration of the task. Tasks are created by the planners much as they would create a list of work items; as they think of something that needs to be done, they create a task card. Card creators are encouraged to identify tasks and not be concerned with the arrangement of tasks into networks. Brainstorming and creative thinking become a standard part of the planning process.

After completing a stack of cards for a given branch on the hierarchical tree (this concept is explained in the next section), it becomes possible to arrange the stack in network format by building the actual network on a tabletop. People need no training whatsoever to do this. Placement, arrangement, and rearrangement continue until the network modeler is satisfied with the work output. Frequently, this process results in the generation of new task cards as missing pieces of information are identified. Identification of resource assignment problems is another outgrowth from this activity. Once the creator of the network is satisfied with its arrangement, predecessor and successor information is written down on each card. Milestone information is recorded to ensure that life-cycle timing remains intact. A comment field is provided to note anything about the task that needs to be remembered.

My first use of the 3 × 5 card system resulted in the generation of over 500 cards by my enthusiastic modeling team. To view the entire network, I brought the cards home and arranged them on the floor. This enabled visual comparison of the work produced by one group with that of another. Analysis and critique of the network model began. I became very critical of serial placement of tasks and questioned each of these arrangements. As I studied the overall network, I took notes, thought about what I was seeing, and became familiar with the work that my team of planners produced. Having completed the arrangement and analysis, I was now in the position to go back to individual planners and make a few informed suggestions while coaching them in areas where it was clear that skills needed development.

This entire process involved over 60 people and was completed in less than seven working days! I was witness to the highest modeling productivity ever achieved by my group. The adjustments were made and the entire plan was now ready for input into a computerized system.

The system of 3 × 5 cards demonstrated its effectiveness, but areas for improve-

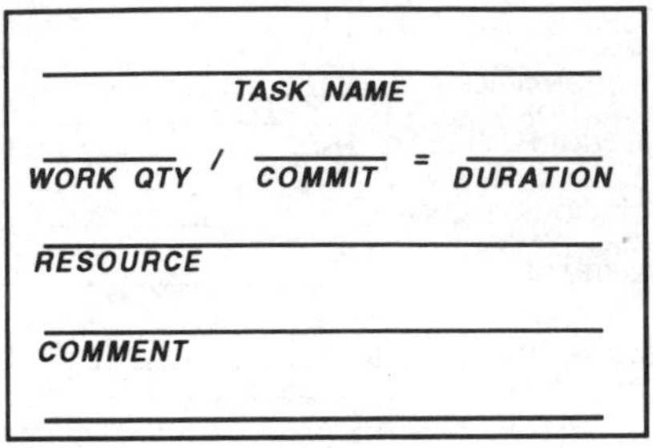

Figure 4.4   PLASTK tag example.

ment also surfaced. The most serious problem was in the duration field. I had asked for one number, but really needed three: work amount, commitment to do the work, and the calculated duration from the previous two numbers. The cards could have been smaller. Each card required a fair amount of text that many individuals found too cumbersome and redundant. Writing down predecessor and successor task names was inconvenient. Layout and visualization of the network required a lot of space. To view networks on a vertical surface required that the cards be taped to the surface. Storing the cards arranged in network format was a difficult task. Computer entry was done one card at a time and did not provide visual network feedback to the computer operator. To overcome these limitations, the concepts behind the 3 × 5 card system were further improved. These improvements are now inherent in the plan synthesis tool kit (PLASTK), which is described below.

The preceding discussion introduced the concepts of project modeling as complexity progressed from the development of personal individual plans at the university level to the on-the-job creation of models for entire cross-functional work groups. In the next section, each step of the process is examined in more detail with focus on the plan synthesis tool kit.

PLASTK is a tool you build yourself. PLASTK, like RA, consists of Post-it or other such tags and layout sheets. An example of a PLASTK tag is shown in Figure 4.4. The fields on each tag, which can be printed up, contain all the information necessary to model a project. Use of this tag in conjunction with the techniques described in future paragraphs eliminates all the problems associated with the application of 3 × 5 inch cards. The technique allows individuals to model their portion of the project in the most naturally intuitive, easily modifiable form available. The tags are used to denote events, milestones, tasks, and activities—all the things that need to go on in the project. The layout sheets help arrange the events and activities in a logical sequence. In addition, each sheet of paper will act as a "limb" of the project hierarchy "tree," containing just those items that belong at that level. Use of the tool makes the development of a project model a fun experience by keeping the process visual and making it easy to change previous work as new information is added to the model.

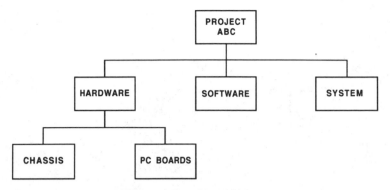

Figure 4.5   Hierarchical tree.

4.3.1   Hierarchical Tree

The hierarchical tree concept is depicted in Figures 4.5 and 4.6. The development of a logical hierarchy structure may seem like a tedious process, especially for the first effort. Many question arise: What is the best way to divide up the project? How many levels should the hierarchy have? Is all this really necessary? Let's answer these questions in reverse order.

As to whether it's really necessary to develop a hierarchy, the answer depends on project size. For a small project, consisting of 60 or fewer activities, model developers could probably manage just fine with all the activities at the top level. For an intermediate size project of 300 to 600 activities, managing all these activities on the same level would be a nightmare! The more time spent up front developing a structured, logical hierarchy, the more easily the project is to model by the cross-functional team.

How many levels the hierarchy should have depends on project complexity, but should probably not need to exceed three to four layers. Some of the work tree "limbs" will have natural subdivisions that will require another level of hierarchical decomposition. For the most part, the shallower, the better. Most medium-sized projects (100 to 500 activities) can probably use two to three layers.

How to divide the work structure will also depend on the particular project. Some activities form convenient groupings according to different aspects of the project. For a simple software project, "Interface Design," "Kernel Design," "Duplication and Distribution," and "Marketing" might be useful subdivisions. A more complex project involving hardware, software, and system integration is illustrated in Figure 4.6.

In this figure, the new product development effort is decomposed by function. This hierarchical structure is identical to the horizontal fields of the new product development process (NPDP) matrix given in Figure 1.11 and explained in detail in the Appendix. In this particular case, the figure shows a hierarchy with three levels. Project XYZ occupies level 1. Program management, research and development,

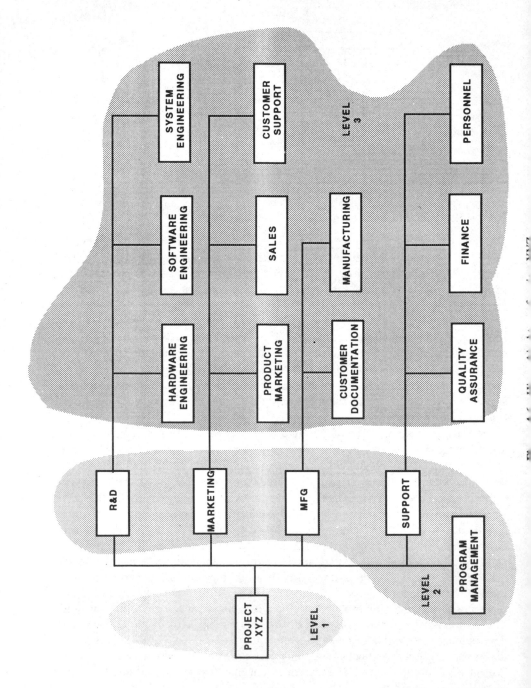

150

marketing, manufacturing, and support reside at the second level. Of the five, only program management is not further disaggregated. Each of the other functional groups is further segmented at the third level of the hierarchy. The hierarchical depth of the horizontal axis of the NPDP matrix is now fully visible. The decomposition of projects by function has been implied since the beginning of the text. At this point in the process, the need becomes more visible than in the past.

It is never advantageous to structure the project according to life-cycle phases. Thus, the hierarchy should not contain blocks showing "Definition Phase," "Planning Phase," or "Execution Phase." Typically, each area concerned with the execution of the project would have activities in each phase of the project, and there would be nothing gained in modeling the project along life-cycle phases.

Two things to remember in deciding the project's functional structure: Model developers will want to maximize the intrapage connection of activities and minimize the interpage connections; and the first stab at deciding the structure will probably not be the last. As the definition of the activities that go into the project takes place, patterns and groups of activities that form a natural proximity begin to emerge. Model developers will become aware of these natural groupings as the project model develops, and should be willing to restructure the project accordingly. The functional decomposition of the organization and that of the project are usually close in structure, but not exactly the same. The tags make restructuring easy. After an organization develops its first hierarchical decomposition by function, only slight modification is needed to reapply the structure on subsequent projects. By the time the second or third iteration takes place, the hierarchical tree used for modeling and the functions identified on the NPDP matrix become one and the same.

4.3.2   Milestone Sequence

Within this discussion, *events* or *milestones* are terms used more or less interchangeably. They are occurrences within the project that take no time and use no resources. "Project Start," "Foundation Complete," "Ready for Test," and "Project Done" are all examples of events. They are characterized as being milestones in time with easily discernible characteristics. The availability of key deliverables is also modeled by events.

Activities or *tasks* take a finite amount of time, and consume resources. They represent the actual work that must go on in order to bring the project to a timely and fruitful completion. "Obtain Feedback," "Lay Foundation," and "Test Code" are examples of activities.

Once the project's hierarchy (or first version thereof) has been determined, the cross-functional planning team begins to create and paste the tags representing the project's milestone sequence on the layout sheets. Program management provides an overview of the project by placing the major milestones and deliverables on the top-level layout sheet. In this example, a milestone sequence is established at level one of project XYZ by the program manager. Once established, this milestone sequence resides within each branch of the tree; that is, all functions structure their work effort to the same milestone sequence throughout the entire hierarchical tree.

A major milestone in this context would be any major event or checkpoint of interest to the highest level of management involved with this project.

As already mentioned, "Start" and "Finish" would be two obvious milestones appropriate to any project. "Finish" might also signal the availability of the major project deliverable. The new product development process (NPDP) matrix time axis provides the necessary link to specifying the milestone sequence. For the NPDP matrix utilized in this text, the sequence consists of a milestone at the beginning and end of each phase. Thus, the definition phase milestones are Begin Definition (Begin Dfn) and End Definition (End Dfn); planning phase milestones are Begin Plan and End Plan; and execution phase milestones are Begin Execution (Begin-Exec) and End Execution (End Exec). Inspection of Figure 4.7 shows how the milestone sequence appears at the top level of the project plan. If your organization has a standard new product development life cycle, this is the appropriate time to reconcile life cycle, milestone sequence, and NPDP matrix time axis symbology—they all should be the same.

To model the milestone sequence, a tag is completed for each event, filling in the name of the event, and 0 for the duration. The "Comment" area is for a more complete description of the event. All other fields remain blank on the tags that represent milestones. On the layout sheet, the horizontal axis is a free-form time scale. For the first layout sheet, which represents the top level of the hierarchy, the one sheet could represent the duration of the project. It is helpful to place the most major events near the top, so place the "Start" event tag near the top of the layout sheet. The "Finish" tag goes all the way on the right. The remaining milestones are placed at appropriate intervals between "Start" and "Finish."

The horizontal placement of the events is not critical: we don't really know yet when the events will happen. All we need to do at this point is get the sequence in proper order.

If, owing to the size of the project, it becomes obvious that there is not enough room to place all events or activities on one sheet, simply expand the time scale by continuing it on another layout sheet. When placement of all major milestones at the top level is complete, the layout sheet should look something like Figure 4.7.

In the figure, we see that the "Start" event kicks off the beginning of the definition phase. Therefore, we place "Begin Dfn" directly under "Start." We draw an arrow from "Start" to "Begin Dfn" to show the "Start" event drives the beginning of the definition phase. The arrow down from "Begin Dfn" indicates that this event, in turn, is driving some event or activity at a lower level. We can also see from the arrow pointing up under "End Dfn" that that event is being driven by some event or activity at a lower level, namely, the final event or activity in our definition phase.

We can also see that our planning phase can overlap the definition phase somewhat. Because all the events at the top level need to be driven by events and activities at the lower levels, the "Begin Plan" event is driven by an event called "Ready to Plan," which, in turn, is driven by the finish of some activity at a lower level. Finally, we see that the end of our execution phase determines the end of our project.

Continuing with the example of our XYZ project, we now wish to map the second level of the hierarchy onto "second-level" sheets, which will consist of the

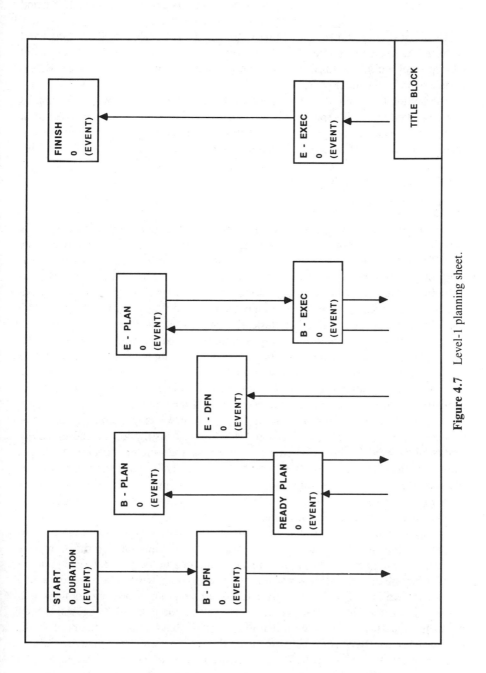

Figure 4.7 Level-1 planning sheet.

153

R&D, marketing, manufacturing, and support boxes. Since each of these functions is also interested in seeing the life-cycle phases, we repeat these events at the second level, renaming them slightly as appropriate for that level.

Thus, for the R&D manager, we have "Begin R&D Dfn," "End R&D Dfn," and so forth. As shown in Figure 4.8, these events are, in turn, connected to events at the top level and at the third level (where the actual activity detail reside). Directional arrows show the flow of events.

To map the third and any subsequent levels of the hierarchy, a process similar to that described above is repeated. The major events and milestones are repeated at each level, renamed as appropriate for that level. In our XYZ project example, the manager in charge of system integration would see a layout sheet similar to Figure 4.9.

4.3.3   Task Lists

Having produced the project's hierarchical tree and the milestone sequence that is embedded within each branch of the tree, the planning team is now ready to start defining and placing activities on its layout sheets. In effect, the cross-functional team of planners begins a distributed brainstorming exercise. The easiest way to accomplish this feat is to have each of the teams do a task listing for its respective cell of the hierarchy. For a team structure such as that illustrated in Figure 4.6, 12 teams would begin to think of the tasks they have to do and create a tag for each task. The "task name" field on the tag is completed and the tag gets placed on the layout sheet. From this point forward, each team contributes to the full realization of the project's model. The teams are now in an excellent position to begin the development of a shared and distributed plan. To initiate the process, an energetic program manager prepares a layout sheet for each planning team and invites team leaders to a kickoff meeting. During this meeting, this manager explains the entire process outlined in Figure 4.1 and provides detailed instruction on the modeling branch of the model. Those invited to the meeting leave with the planning sheet they must complete and return over the next few days. Thus, each team manager or leader leaves the meeting with a sheet that looks like Figure 4.9 and with a pad of tags. Each manager along with his or her team then begins the process of task creation and placement on the layout sheet.

Instructions on how to use the layout sheets should be given. The PLASTK system makes this simple. Modelers are informed that all they have to do is create a task tag for each activity they want to have contained in their model. The creation process can be done individually and by groups. Having some time for individuals to prepare preliminary task tags prior to attending group modeling sessions is a definite time-saver. To prevent planners from getting too detailed, suggest that they try to structure task definitions so that they contain about 80 to 160 hours of work, but make it clear that it is far better to err on the side of more detail than too little.

The creation process results in the generation of many task tags attached to the group's layout sheet. Up to this point, the tags are not arranged in any particular order: they are randomly distributed over the layout sheet. Resourceful teams may

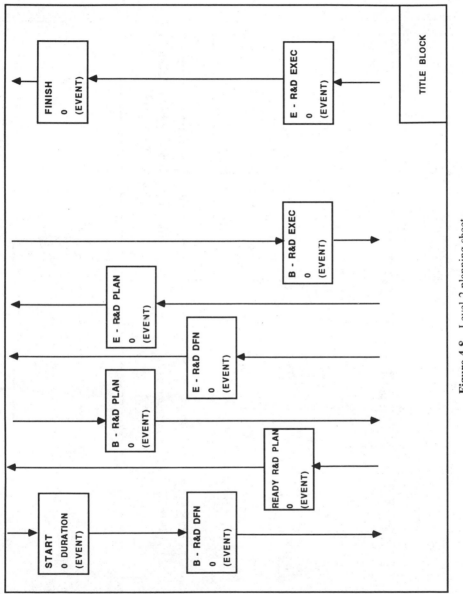

Figure 4.8 Level-2 planning sheet.

155

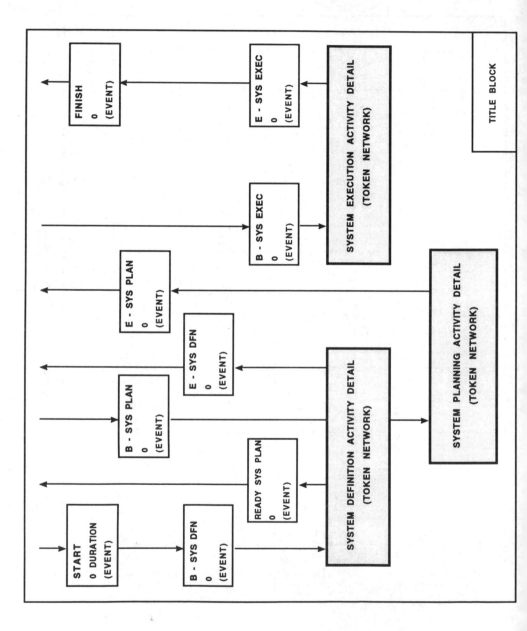

156

have already begun the process of placing tags within particular time segments represented by the milestone sequence. Taking this step is convenient but not necessary—the planning team will eventually group items being performed in common
segments of time. Hierarchically, like items tend to be sorted by time and life-cycle
position. Time sorting is rather generic. Arrange the task tags in a list that best
places each task in approximate chronological order. This sorting results in the familiar "to do" list representation for the entire task list. Since milestones were defined in the previous step, the chronological list is segmented into smaller lists that
fit within the established milestone sequence. One main list is now broken down
into a number of items that must be completed before a milestone can be achieved.
Lists of hundreds of items can be time-sorted and broken down using this technique
in a very short period of time by individuals and groups with amazingly accurate
results. Frequently, the sorting and breakdown will result in changing existing task
tags and identifying the need for additional task information.

Each modeling team within the project hierarchy follows a similar process; as a
result, expect the modeling team to generate a relatively large number of task tags.
If our example project XYZ were being modeled by the 12 lowest-level groups
shown in Figure 4.6, it is easy to see how 500 or more task tags might be created
for the project.

With tasks placed on the planning sheet, attention shifts to the next field on
the tag.

4.3.4   Task Work Quantity Estimates

The people that created the task lists for their function are in the best position to
determine the work quantity contained within each of the tasks. Where possible,
they should draw on historical data for similar tasks to make the work quantity
estimate. Reliance on historical data becomes possible only when a database exists.
Those who follow the concepts contained in this text on a subsequent project will
have a database of task work quantity information available. Their estimates based
on history will improve in accuracy as estimates are verified and measured under
actual execution conditions.

For those working without meaningful data to make a task work quantity estimate, it becomes necessary to produce numbers that are realistic for the task under
consideration. Where possible, these numbers should be derived from known information. Estimates produced by inference are usually more accurate than those that
are just best guesses. Sometimes solicitation of input from those who have done
similar work in the past helps in making the estimate when historical data is not
available. This is one way to access the database of human experience; however, be
warned that people tend to remember the best of the past and forget the painful
delays. People also tend to confuse the concepts of work quantity and duration.

The main question that has to be answered is, How much work is there within
this task? Is the work quantity 10 hours, 40 hours, or 160 hours? It's generally much
easier to determine an accurate work quantity number than it is to specify a figure
for task duration. Of most importance is to never ask the question, How long will it

take? The answer can be computed from knowing work quantity and assigning a resource to work on the task at a given rate. A 20-hour work quantity task takes 20 days to complete if worked on for one hour a day, but could get done in two days if commitments are made to work on it at a rate of 10 hours per day. Well over 80 percent of the work that goes into the development of a new product can be modulated in this fashion. This fact will be exploited within the computer system to adjust project timing as a function of resource commitment to task. For now, planners only have to determine a single work quantity number for each task. Some of the people making task estimates will require coaching. They will ask you questions like, "How detailed should we get? I can give you 20 activities each containing four hours of work, or one activity containing 80 hours of work!" A general guideline is 80 to 160 hours of work quantity at this level of project modeling. Individuals and managers of the various functional groups may want to have more resolution at the local level. At the project level, too high a resolution may make cross-functional plan integration and tracking impossible. Clearly, the work quantity estimation process is likely to impact the structure of the task lists. Where necessary, make the adjustments to the lists to get convergence between task descriptions and work quantity. Where the work quantity of a task is longer than 160 hours, work to decompose it into smaller task segments. For those tasks that have a work quantity that is less than 80 hours, work to combine similar elements to approach the guideline. Exceptions do result. Where combination does not make sense, go with the original structure. Decomposition to 160 hours, or less, should be achieved.

Another frequently asked question is, "Should I provide best-case, average, and worst-case work quantity estimates?" The answer is yes if you have a historical database that statistically produces these numbers. If data do not exist to compute the best-case, average, and worst-case numbers, do not try to guess what they might be—you'll probably be wrong. Instead, use your best information and judgment to come up with the best single number for each task that you can. The system you are working within will be averaging many inputs to arrive at a final answer. This answer will be compared to estimates made in the definition phase to elevate awareness of where differences exist. Comparison, averaging, reconciliation of differences, and plain good judgment are better than three arbitrary numerical guesses.

4.3.5   Resource Assignment to Task

Individuals and teams doing the modeling to this point will have some idea of who they would like to assign to a particular task. However, it is usually not within their power to authorize and actually make the assignment. As such, space on the task tag is provided for the planners to suggest the name of the individual they would like to have work on the task. Knowing that resource information at this point in the process is a suggestion only makes it possible to think more idealistically. While making the resource suggestion, it is appropriate for the planning team members to make a preliminary commitment of the suggested person's time. Once this is done, simple mathematics is used to compute task duration and then fill it in on the task tag. Managers who have some ideas and preliminary thoughts on how resources are likely to be allocated can use their intuitive skills to help guide the resource assignment and commitment to task process. Obviously, the more accurate the data at this point in the process, the less likely the need for detailed manipulation and processing once the information resides within the computer.

Anyone that has made the attempt to assign resources to tasks, without the benefit of having the information from the definition phase and without some managerial guidance on which resources are likely to be available to the project/program, knows that it is an exercise in futility. With this information available, people are working with quality data to produce resource and task links that are, for the first attempt, realistic. Some changes to the results are inevitable as more detail is added to the plan from other people contributing to the development of the product. This change, however, can be expected to be minimal and meaningful. The tools that will be used to make the adjustments will be set up with good data on which to base future resource allocation decisions.

It is important to have a person's name associated with each activity. Activities without a person's name typically don't get done! If the activity is too far in the future for the modeler to come up with a name yet, the modeler should be encouraged to add his or her own name to this task until suitable resources have been identified. Make sure the modeler understands that *until he/she gets someone else assigned to the task, the implication is that he/she is the one responsible for that activity.* Frequently, a first-pass resource assignment will include the name of the functional manager who is responsible for actually committing the resource to the task. This is appropriate. Once input is machine processed, this manager will then get feedback on all tasks that are linked to his or her name. If the list is too long, it's likely to capture this person's attention!

4.3.6   Task Networks

Task networks are readily constructed from the completed task tags that occupy a given time bin associated with each milestone. Building a network becomes a natural result of the PLASTK process. All that is needed is to arrange the task tags in network format and draw arrows showing their connectivity. When coaching people in the network arrangement of activities on the layout sheets, one concept that needs to be made very clear, and one that is not intuitively obvious to most people, is how to designate activity dependencies. The most common form of dependency is the finish-to-start dependency. This implies that activity B *cannot* begin until activity A is completed, just as the framing cannot begin until the foundation is laid.

Task network models are formed by connecting tasks to one another in series, parallel, or a combination of the two. The most important item to remember about building the network model is to make network connections based only on the consideration of how the tasks relate to one another. Use a series connection only if activity B cannot start until activity A is finished. Use parallel arrangements in all other cases. Question all series connections and look for ways to remove them or decompose them into short work content structures. If activity B can't start until activity A is finished simply because the assumption is made that the same resource is working on both, but by the nature of the work involved the activities could both actually start at the same time, this is not a true finish-to-start dependency, and shouldn't be modeled that way. The computer tool being used will be able to handle the resource conflict on its own. From a time-to-market perspective, serial relationships slow down the new product development process. A network that consists of only essential serial connections is referred to as a *minimal serial network*.

The production of a minimal serial network *now* greatly simplifies using the computerized resource allocation system in the next step of the process. With the minimal serial network in place, it becomes possible to determine time to market as a function of the resources that are applied to the product's realization. To do this, resources are assigned to each activity and committed to work on the assigned activity at a given rate. The computer tool then computes the task duration from the information available and begins to provide detailed feedback about the effort's TTM. After the commitment of all available resources, leveling and constraining algorithms are invoked to ensure the resource commitments made are held within the daily limits. More on this topic in Chapter 5.

A finish-to-start dependency is shown with an arrow drawn from the right side of an activity to the left side of the dependent activity. Another connection type is start-to-start with an optional lag. This implies that activity B cannot start until activity A has also started, usually with some lag time in between. Thus, we don't really have to wait until the trench is finished before we start laying the pipe; we can start laying pipe perhaps three days after the trench is begun. This is a start-to-start connection with a three-day lag.

Usually, when there is a start-to-start connection, there is also a finish-to-finish connection (with optional lag) as well. Thus, we cannot finish laying the pipe until three days after finishing the trench. Assume task B below task A. A start-to-start connection with lag is shown by drawing an arrow from the left side of activity A to the left side of activity B, with the lag next to the connection. Finish-to-finish with lag is drawn from the right side of activity A to the right side of activity B, with the lag written in next to the arrow:

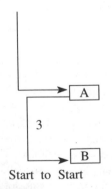

Start to Start

Events that start a phase should be connected to the first activity in that phase. The last activity of that phase should be connected to the event that signals the end of that phase. Any intermediate events can be connected in the same way; just be careful to show the directional arrows to indicate the flow of work:

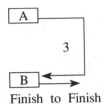

Finish to Finish

Utilization of PLASTK to model the project helps the entire modeling team prepare a shared plan for the project. The team will have produced layout sheets whose number is proportional to the complexity of the project being modeled:

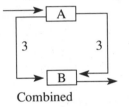

Combined

Probably, the team will have generated a task tag count that is in the high hundreds to low thousands. Collectively, the team will have produced a model of project reality that can service key decision-making information needs well into the future. Information from the resource allocation arm and the project modeling arm of Figure 4.1 is now ready for input into the computer tools. Additionally, the format of the output from both "arms" is compatible with the input requirements of quality project management software tools. If your organization is fortunate enough to have decided on the right tool for the application, the program manager can hand the network sheets to his or her data input assistant for transformation into the electronic format. If your organization is not working with the best computer tools, the next chapter begins with a framework to help make the necessary hardware and software selection decisions in a logical fashion.

5

PLAN ANALYSIS

In this chapter, you will learn about

A procedure to select optimal plan analysis tools.

Tools and techniques to analyze project plans fully.

Use of the KMET chart in plan analysis.

Reconciliation of planning phase targets with definition phase estimates.

The production of computer-generated "to do" lists.

Problems that can be solved or minimized by the application of good planning techniques.

The contributions of the planning phase to the overall process.

Making the *gravis arbitrium*.

Force competitors to play catch up!

BMC Software Inc.

5.1 INTRODUCTION

Computer tools have had some impact on the new product development process up to this point, but in most areas the process and tools have not had a tight coupling. Chapter 5 begins the integation of planning, tracking, analysis, and reporting tools into the overall process. The remaining topics of database creation, data analysis, comparison of definition estimates with planning targets, and generation of reports and charts identified in Figure 4.1 are discussed in detail. Initially, the data developed in Chapter 4 is used to evaluate tool alternatives. Based on the evaluation, a tool is selected as the Chapter 4 data are fully applied and analyzed. Information is viewed from a number of perspectives to provide feedback to the project and management team about key project attributes. Horizontal and vertical new product development process participants are introduced to the KMET chart, individual computer-generated "to do" lists, the scheduled completion report, and the scheduled exception report. Once plan analysis is complete, the planning team will be instructed to compare the results of the planning effort to the estimates produced during the definition phase. Where significant differences exist, the process of reconciliation is initiated until satisfactory results are produced. After becoming acquainted with the various plan analysis techniques, the management and project teams will know when they have produced a plan that contains sufficient quality information to begin the execution phase.

Discussion is now directed to the evaluation and eventual selection of computer tools that will be used to support the new product development process.

5.2 COMPUTER TOOLS

Figure 4.1 helps in visualizing the contribution the tools are expected to make to the planning process. Key new product development process managers are now in a position to take the project model developed using the tools and techniques described in Chapter 4 and use it as an evaluation mechanism for the selection of computer tools and software. Your organization will find this approach to be far superior to the often-followed alternative of selecting tools first and then trying to develop models that satisfy tool requirements. A list of important tool selection parameters to consider follows:

• Supports hierarchical tree planning techniques.

The plan synthesis tool kit (PLASTK) deals with hierarchical decomposition of the project by utilizing a full planning sheet for each branch of the hierarchical tree. For effective results, you will want the tools and software to be capable of working with this type of project decomposition.

Software manufacturers frequently employ two techniques (actual hierarchical trees and outlines) to support hierarchical planning. Both are effective and selection is a matter of personal preference; however, of the two, an actual tree structure provides the best visualization of the project's structure and makes it extremely easy to move about within the model.

• Functions effectively on >1000 tasks.

The key word above is *effectively.* A typical application of PLASTK will result in the generation of a model that consists of hundreds of tasks and milestones. It becomes easy to visualize how the initial model could grow in complexity as more becomes known about the project. Project/program modeling software must be capable of handling a large number of tasks on the initial effort and provide the reserve to meet projected future needs. One thousand is a modest number for many projects. Whatever the task and milestone count of your initial model, allowing for two to three times the amount is suggested. Knowing this expanded number can expedite your decision process.

• Can be used by horizontal and vertical decision makers.

Your model was developed by a horizontal and vertical team. Logically, you would want those who contributed to model development to be capable of using the model as a decision-making tool once it is available in electronic form. As a result, the hardware and software selected should be such that it can be made available to the entire decision-making team. Network implementations are frequently of great value where project teams are separated physically and geographically. Access to the same project database gives decision makers the capability to assess cause-and-effect relationships of various alternatives once the project is under way and as adjustments are made to the plan.

- Can be used by support personnel.

Whenever possible, model developers should get help in having the information generated using PLASTK input into the computer system. Model developers tend to shy away from large data-input responsibilities. Data-processing support tends to keep the developers focused on doing a quality job creating the model rather than on minimizing work for themselves. One person trained on data input from PLASTK can support many projects within an organization provided the person finds the computer tools involved easy to use.

- Supports project tracking by storing a baseline reference.

A model is put into the computer system, analyzed, and then refined by the individuals that produced the initial model. Shortly thereafter, the final electronic representation of the model is produced. This first electronic model becomes the reference for the execution phase of the project/program. It is significant in that it represents the best effort that the planning team could produce. An effective computer tool should be used to keep track of the planned dates and actual dates for each activity, as well as the original plan for comparison purposes.

Future work will be compared to this initial model to give the creators of the plan the feedback they need to keep the project/program on track and to provide them with information that will help them become better modelers in the future. It is not likely that you would choose a tool that does not provide this capability.

- Supports merging of projects/subprojects.

Hierarchical modeling and this tool attribute go hand in hand. Model development using PLASTK can be taking place concurrently in research and development, marketing, manufacturing, and support areas. Each area can then input the model they developed and create a data file of their own for analysis, review, and revision. Once each group is satisfied, each of the independent files can be linked, merged, and connected to the other files.

In this way, the computer tools become the focal point for model integration. Being able to merge projects/subprojects makes integration easier while it enhances the ability to analyze the combined plan.

• Does calculations fast.

PLASTK simplifies the evaluation of various software alternatives on this important parameter. You can take the model that was developed, put it into the computer system, and measure actual combined hardware and software performance.

Speed of computation is important when working with models that contain a large number of elements. It is not unusual to see execution speeds that differ by over 10 to 1 from one brand of software to another. Quality software can fully analyze a 1000-element model in less than 30 seconds when running on an industry standard computer with a clock frequency of 16 MHz. Higher-performance hardware makes it possible to grow model complexity and still keep calculation times well within an acceptable recalculation time window.

• Supplies the ability to export data.

No brand of project management software on the market has all of the analysis and graphics features needed to analyze and display the information content that makes up the model accurately and thoroughly. To do more with the data using other programs, it is necessary to rely on the data export feature.

Later in this chapter, a software package called plan analysis software (PAS) that utilizes the export feature will be described. With PAS , you can produce computer-generated "to do" lists, generate KMET charts, list all tasks alphabetically or by scheduled completion date, and generate a number of other key analysis reports. Besides using PAS, you may want to do an analysis or presentation of your own using spreadsheet, statistical, or graphics programs. The export feature makes it possible to extend the utilization of the information contained within the project management software database.

• Uses laser printer.

As you begin to work with large models, you will soon discover that you will no longer want to work with plotters and the output they produce. The plotters are just too slow and the presentation space requirements for the output they produce is too extensive. You will find yourself working in a dimension that is beyond PERT network charts and Gantt charts. The days of covering walls with D- and E-size plotter output are history. Rapidly generated output from a laser printer will become the official written documents utilized by the project/program.

Modelers and planners no longer have the luxury to wait for days to get feedback on what is in the project database. The necessary documents to run a project in the fully modeled form are individual "to do" lists, scheduled completion task list sorts, alphabetical task list sorts, and KMET charts. The laser can produce these documents in a matter of minutes for a 1000-task model, whereas, a plotter could require up to two days to produce conventional output.

- Supports graphical input of networks.

Simply stated, a graphical interface is better than a tabular one to obtain the most accuracy and build confidence in synthesizing the network. Moreover, a graphical interface is fully compatible with the output produced by PLASTK. If you have any doubts, try describing a network using a table or outline format and then try doing the same thing with a sketch of the network. In which case is it easier to spot an error? In which case is it easier to make an error?

- Supports table entry of task data.

While the graphical input of networks enhances user confidence and accuracy of the network, there are times when a more rapid method of data entry is helpful. The initial stages of massive data entry is such a time, when the data entry might be delegated to support personnel. A tool that supports the table entry of data speeds the mass entry of amount of work, commitment, resource, activity code, and other related task data.

- Calculates task durations from work quantity and resource commitment.

All useful software tools allow the user to enter estimated task durations directly; that is, task A will take 10 days to accomplish. This method is suitable for some initial estimates. A more sophisticated approach is to enter the amount of work required (task A requires 80 person-hours) together with an estimate of resource commitment (there is expected to be one engineer available for eight hours a day). The software tool then makes the calculation: 80 person-hours ÷ 8 hours per day = 10 days' duration. This capability permits significantly more flexibility to apply resources to the effort to realize specific market-entry goals.

- Performs resource leveling and constraining.

This functionality, when intelligently implemented, can take a lot of work out of resource management. Above, "leveling" means the ability to smooth the peaks and valleys of resource usage as much as possible, within the confines of available float; that is, tasks are moved but no activities on the critical path will be altered. In this mode, project milestones set by the critical path will not move in time. "Constraining" means pushing whatever activities are necessary to achieve a smooth resource usage. Application of the constraining algorithm usually results in the end date slipping.

Working with the output from RA and PLASTK to evaluate hardware and software to the previously stated criteria should lead to a final computer tool decision. Knowing what to look for and knowing your particular situation will lead to a decision that is quickly produced. It will be a decision that has team support and commitment because many of the modelers will have been involved in helping make the tool selection from a set of known selection parameters.

The section titled "Software Evaluation" in the Appendix is provided to simplify the tool selection decision process further by establishing an initial point of reference for your comparative evaluation. This evaluation, one of many performed by a business associate, Ray Hopper, is based on the previously described evaluation parameters. You can see how one software package performed and compare your test results to those described in this section of the Appendix. Using a relative decision-making process, a clear choice for your application should emerge in a short period of time.

Future discussion assumes that the tool decision has now been made and that the tool is available horizontally and vertically throughout the organization.

5.3 DATABASE CREATION

Creation of the project's database is a straightforward process. All that needs to be done is to duplicate the information contained on the PLASTK network sheets within the computer system and add the resource and calendar information from the left path of the planning process flow diagram shown in Figure 4.1. Those of you who took the steps to select a tool obviously will not want to repeat the reentry of data from PLASTK. Where possible, each functional area (research and development, marketing, manufacturing, sales, etc.) should be encouraged to input the information they created. Ideally, modelers from each functional area should have tools and data input support available within their work areas. Where this is not practical, consideration should be given to having central resources available to assist with the input process. Once the data from each functional area are installed within the system, the project/program manager and his or her support staff can begin the process of joining the data files and adding the network connections between the files.

To expedite the above process, it is a good idea to make a template for the project before beginning to input the data. The template can be distributed to all functional groups to help them input their task data. The template consists of the project's hierarchy and milestone sequence but not any task data. In effect, it is the complete matrix and network representation of the project without specific task data. Remember, for each branch of the tree a milestone sequence was established and appropriately named. This sequence appears at all levels of the hierarchy. Levels of the hierarchy are vertically linked by arrows; milestones within each branch are connected to the appropriate milestones within other levels. The entire vertical linking between levels of the hierarchy is created in advance of entering task and network data. You must remember to link the events between levels in the hierarchy in the correct order. At this point, predecessor and successor relationships (to/from direction of network arrows) must be accurately established. The same structure that was built in the section labeled "Milestone Sequence" described in section 4.3.2 is duplicated electronically before beginning to input task and resource data. Once complete, the template serves the needs of the present project and establishes a reusable structure for future projects. You will also find the template to be a useful tool when integrating inputs from a number of independent functional areas.

Assuming that a template exists, the only remaining preparation work that needs to be performed is to input the project's resource list, resource costs, and calendar. The data for these fields was created using RA (resource allocator), accounting information, and simple work scheduling information. Most project management software requires that that resource and cost data be entered in tables. Timing information is generally entered through a visual calendar structure in which global daily work hours, weekly workdays, and holidays are established for the overall resource pool.

Now the task of entering the networks for each branch of the hierarchical tree begins. If you followed instructions correctly, you will find that the network detail appears at the lowest levels of the hierarchy. A tool with graphical network entry allows the input person to copy the network directly from the PLASTK sheet and produce a similar visual structure. A tool with outline entry requires that you produce a PERT chart to view the network after data entry is complete. Tasks need to be linked to each other and to the appropriate milestones within their respective hierarchical branch. Occasionally, it will become necessary to link information flow between different branches within the hierarchical tree. At all times, care must be taken to establish the correct predecessor/successor relationships between tasks and their linking to others tasks in differing branches of the hierarchical tree. Being able to compare the information within the computer system to that on the PLASTK sheets has a significant effect on error reduction; computer-generated networks should have the same structure as the manually generated reference sheets. Differences in network structure need to be resolved to complete this segment of input work.

Lastly, the data available about each task is entered. Again, the data is taken directly from the previously described task tags. Resource assignment to task, task work quantity, work commitment, and task notes are the only information that remains to be input. Typically, the input process runs at a productivity rate of 20 to 30 tasks/events per hour for an experienced input specialist. This figure can be used to estimate the time needed to perform the input operation based on the number of input personnel available to do the job. Once complete, the electronic version of the project's model will provide a new insight into the infrastructure of the remaining portion of the new product development process. Analysis of this database will initiate the learning and growth process for the entire new product development team.

5.4   DATA ANALYSIS

Now that the data for the entire project are resident within the computer system, it is possible to begin a systematic review of the information.

First, the project's database is evaluated from a network and content perspective.

Does the network within the system agree with the PLASTK network? Are there more serial connections that can be eliminated? Is the network truly of the minimal serial variety? Once you have produced a network with a minimum of serial connections and you are convinced its structure is correct, you can then begin working with and checking the information associated with each task. Is the necessary task information in place? What is missing? Make certain that work content and resource commitment to tasks are being used to calculate activity duration. For tasks where duration is specified directly, determine how the needed information to produce a calculated duration can be obtained and then add this data to the project's database.

Second, begin reviewing the data on each task from a resource selection, resource commitment, work quantity, and duration perspective. Is the resource assigned to the task a good choice? Is the resource assigned to the task actually assigned to the project? Closure between resource assignment to project and resource assignment to task receives its first critical test at this point in the analysis. When using PLASTK, it was not easy to check each resource assigned to a task with those assigned to the project using RA. The computer tool makes this test automatic. All that needs to be done is to compare the names within the resource table used at the project level with the electronic version of the list obtained from RA. Of course, differences must be reconciled. With preliminary resource assignment to project and task issues resolved, attention shifts to review of work quantity, work commitment, and calculated durations for each task. Does the work quantity estimate appear to be reasonable? Is the numerical value in the commitment field correct? For tasks that have only a single resource assigned, are the computed durations about two to four weeks in length? Reviewing the data associated with each task should consume about two minutes, enough time to become comfortable with the information and make suitable adjustments to it. Therefore, review of a 300-element model should take approximately 600 minutes, or 10 hours. As a rule of thumb, give yourself one day to review the information associated with each 250 tasks.

Third, apply the tool's resource loading capabilities to view how each resource within the database is utilized. Which individuals are significantly overdeployed? Which resources are underutilized? As the histogram of hours worked per day for each resource is viewed, areas of concern are certain to be identified. At this point, work to remove peaks and valleys in assignments that are clearly in "gross" error. By my definition, "gross" errors in the resource histogram are those where individuals appear to be working more than 16 hours per day and less than 2 hours per day. Think through the means by which loading can be adjusted to fall within the 2- to 16-hour range for each resource. Once acceptable alternatives have been determined, make these course adjustments to the database. Please note that up to this point nothing has been done to adjust the schedule of the project, but we have now completed all the preliminary work necessary to begin the schedule adjustment process.

Having the data assembled about the new product development effort in this format now makes it possible to determine schedule and time to market (TTM) as a function of the resources that are applied to the product's realization. To exploit the capability, you adjust the assignment of resources to each activity and commit the assigned resource to work on the activity at a given rate. The computer tool then computes the task duration from the information available and begins to provide detailed feedback about the effort's TTM. After the commitment of all of your available resources, you'll want to use leveling and constraining algorithms to ensure the resource commitments you have made are within the daily limits typically worked within your organization. You'll find yourself iterating and adjusting resource-related data to get what looks like the best overall balance between the resources available to the project and the project's schedule.

Frequently, after consuming all of your available resources, you will still be separated by some distance in time from meeting your TTM objectives. If front-end work has been done correctly, the exercise becomes one of adding resources in critical areas until TTM objectives are realized. To modulate resources to achieve project objective will require continued horizontal and vertical interaction within the organization.

Continued review and utilization of the data will rely on analysis features resident in quality project management software products plus utilization of plan analysis software (PAS) tools to provide added, necessary capability. Future discussion expands on the information supplied in the introductory paragraphs of this section. Analysis will first focus on network and time properties. Second, effort will be devoted to examine resource utilization.

We will assume that the execution phase is divided into three parts: prototype (Proto), pilot, and production (Prod). Each phase has a beginning and ending milestone labeled as B-Proto for begin prototype phase, E-Proto for end of prototype phase, B-Pilot for the beginning of the pilot phase, and so on. Additionally, the hierarchical structure given in Figure 4.6 is utilized in this discussion. For each branch on the tree, the major milestone names are modified by the branch name. Thus, for the research and development branch, the major milestone names become B-R&D Proto, E-R&D Proto, B-R&D Pilot, E-R&D Pilot, B-R&D Prod, and E-R&D Prod. This convention is then followed for each of the 15 remaining branches in the hierarchy. Please note that the convention is the same as that provided in the PLASTK discussion in Chapter 4 (see Figs. 4.7–4.9) except detail is removed from the definition and planning phases while more milestones are added within the execution phase. All of the names describing milestones/events and tasks exist in the project's database.

Database analysis starts at the top level of the hierarchy tree by viewing the overall project by each of its major milestones through the execution phase. A milestone summary by phase is shown in Figure 5.1.

The figure shows that the initial plan calls for project completion in five units of time. The prototype phase consumes about 50 percent of the project's time to market. Pilot phase and production phase combined time is approximately 60 percent of the total new product development time. The ratio of prototype time to pilot plus production time is above the 50 percent historical norm for this type of product development effort. Naturally, those doing the analysis want to find the reason.

Another view of the data is assembled by function. This view is shown in Figure 5.2. From this perspective, it is easy to see that manufacturing needs about one

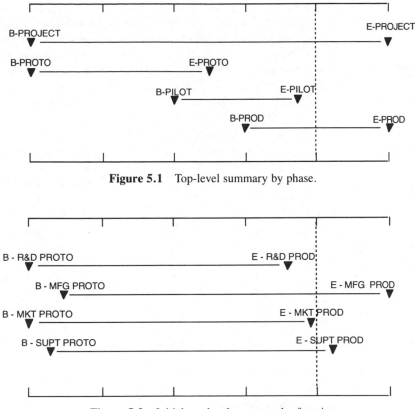

Figure 5.1   Top-level summary by phase.

Figure 5.2   Initial top-level summary by function.

major time unit more than other functions to accomplish its tasks to introduce the new product, whereas the remaining three functions are closely grouped around the fourth time unit. The natural question arises: What would it take to bring all functions into better time alignment around the fourth time unit? Doing so would reduce time to market by 20 percent. What options are available to gain a 20-percent reduction?

The manufacturing team is asked to review the estimates made in the definition phase to analyze this preliminary planning phase result. The complexity of the work, resources available to do the work, and the productivity of the resources must be thoroughly examined. The first question to ask is, Did the resultant plan meet the estimates set at the end of the definition phase? If not, why not? Has work complexity expanded? Are too few resources assigned to do the job in four units of time? Is productivity lower than expected? Careful review of the data at the third level of the project's hierarchy supplied by manufacturing-materials, manufacturing-assembly, and manufacturing test will begin to suggest methods to remedy the situation.

Decision makers need to assemble and decide collectively what is to be done. Often, other members of the planning team need to be involved. On complexity issues, research and development may have to modify the design approach, marketing may be asked to modify certain attributes of the product's definition to get a better time to market, or functional management may have to approve a change in strategy, tactics, or logistics. To solve resource problems, the team may have to look to other functional areas to fill critical needs, evaluate subcontracting some of the work, consider adding new people to the manufacturing staff, or look to outside part-time help to fill critical positions at critical times. Productivity figures should be computed to see if the planners might be "sandbagging" their task targets. If this is the case, you may want to assist those who are having the difficulty by asking them to provide analytic justification for the numbers that were produced. If the targets seem reasonable, consideration should be given to taking actions that will have a direct impact on productivity: skill development, tool additions, adoption of new work strategies, and so forth.

Figure 5.3 shows the results produced by the manufacturing team as it worked with others on the project to revise the plan. What was initially thought to be a minimally serial network was not. The team identified ways to add more parallelism and start working on the critical manufacturing-related tasks at an earlier date. In addition, it identified work quantity, resource, and resource commitment alternatives that reduced manufacturing's time to market by a sufficient amount so that the team no longer caused an obstacle on the critical path. Throughout the revision process, intelligent network modifications were produced and evaluated on the tool. Each change produced in the manufacturing portion of the database was made with complete awareness of impact on other functions while the effect on the overall project schedule was computed. The inner workings and dependency relationships become highly visible to those working with the database—knowledge of how the entire effort functions expands and improves in accuracy with each iteration and subsequent evaluation.

Note that only a portion of the database was revised; the vast majority of information content remained unchanged. Under these conditions, it is usually best to do the "what if" analysis directly on the tool until satisfactory results are achieved. In

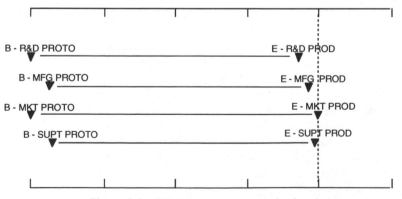

Figure 5.3   Final top-level summary by function.

effect, you return to the synthesis mode and use the tools to help you think about and try various alternatives. Once satisfied with a particular alternative, no further revision needs to be made on this portion of the database.

The same effort that went into reducing manufacturing's time to market must now go into verifying that the targets produced by research and development, marketing, and support are in close agreement with the estimates produced during the definition phase. Again, each function's beginning-to-end plan needs to be carefully examined against previous complexity, resource, and productivity figures. Each functional area needs to convince itself that the database reflects the best that it can do. It's quite likely that as inspection intensifies, the rate of change to the database will increase. You'll have an entire team of modelers and planners doing synthesis on the tool to further improve the accuracy of the information. Of course, only time will tell if it's really right.

As a final test, the database is summarized by function and by phase. This view is shown in Figure 5.4. From this perspective, you can see when milestones will be completed at the upper levels of the hierarchical tree. The prototype phase ends at about the same time in research and development, manufacturing, and support. Marketing finishes earlier and immediately begins working on the pilot phase. To prepare for the pilot phase, manufacturing must start well in advance of the completion of the prototype phase, whereas R&D and support start the pilot phase only slightly ahead of the completion of the prototype phase. As the project enters the production phase, the work performed by R&D is scheduled to complete first, while marketing and manufacturing set the pace for product introduction.

Keep in mind that a network, or collection of networks, lies beneath each of the summary bars shown in Figure 5.4. Hundreds to thousands of pieces of information can be reduced and placed in this format to view the entire project objectively. A data summary of this type is just one of the many benefits of doing a hierarchical project model.

Once plan developers and key members of the new product introduction team are satisfied with the network analysis, the next step is to export the database to the plan analysis software (PAS) tool for performing KMET chart analysis. The discussion that follows is based on a KMET chart analysis for the R&D branch of the hierarchy tree. Data and discussion are adapted from an actual new product development planning phase situation.

The task data for R&D hardware, R&D software, and R&D systems were combined and sorted by scheduled completion date. An accumulation of task completion as a function of time is graphed on the KMET chart. A similar analysis, performed for each branch of the tree, is not given in this discussion to keep attention focused on procedural issues.

The KMET chart is designed to serve three important functions. First, in the planning phase, it provides an excellent measure of the plan structure and detail in a single-page summary. Second, during the execution phase, it tracks key milestones over time to measure slip referenced to the original plan. Third, plan structure is recorded over time to give an indication of the growth and complexity of the project/program. Once familiar with the chart, a user will be able to extract this

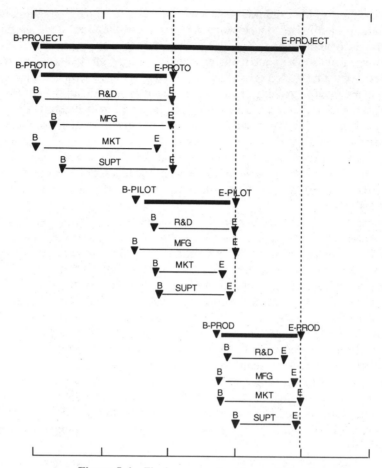

Figure 5.4   Final summary by phase and function.

information at a glance. This section will focus on using the KMET chart during the planning phase. Chapter 6 will deal with KMET chart utilization during the execution phase.

A comparison of two KMET charts (Figs. 5.5 and 5.6) is used to illustrate the first function: the measurement and single-page display of plan structure and network detail. Figure 5.5 shows that the R&D plan for project XYZ consisted of about 63 tasks distributed over a 60 + week period. On the average, the plan was for about one task per week to be completed, but significant departures from a linear running rate are evident throughout the plan. Close examination of the slope of the curve reveals three regions that need further analysis. Note the steep rise in slope just preceding the design review, the flat slope preceding the E-Pilot milestone, and the lack of detail from E-Pilot to E-Prod. Any ideas about the causes of the major changes in slope?

In these abnormal zones, it is possible to return to the network to gain a better understanding of the exact cause; however, a little speculative thinking will prob-

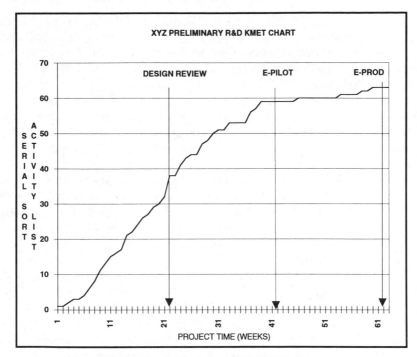

Figure 5.5   Preliminary R&D KMET chart.

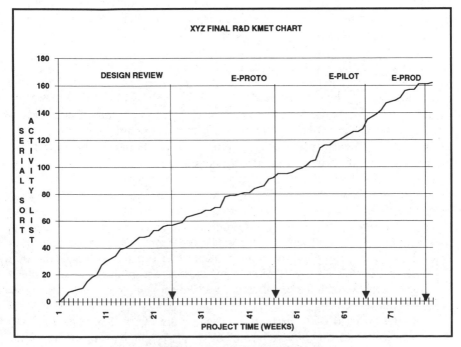

Figure 5.6   Final R&D KMET chart.

ably produce the same results. As might be expected, the slope rise preceding the design review milestone is produced by having work elements come together for the milestone. Note that following the milestone, output goes to zero for a period of two weeks. This plan seems to have a crash into the milestone built in! We have a work group that is planning to work long and hard to get to the milestone and then kick back a bit after it's over. In the vicinity of week 34, output again goes to zero. In this case, the reason is that future work is being delayed by the wait for needed output from another individual. The lack of detail from E-Pilot to E-Prod exists because visibility, of work needing to be performed this far into the future, is "fuzzy."

Now that problems with the plan are identified, it becomes relatively easy to focus on these regions and make the needed improvements. The KMET chart analysis suggests the initial plan is lacking in detail for an 11-person team, contains an improperly structured network around the design review milestone, wastes precious project time preceding the E-Pilot milestone, and lacks sufficient detail from the E-Pilot milestone to completion.

Now knowing where to focus attention, the team involved produced the KMET chart shown in Figure 5.6. In this replanning exercise, detail increased to 162 tasks. With 11 people on the project, this translates to tracking about one task per person per month on the average. Historical network and task work quantity data were used from previous projects completed by this organization. Resource commitments and adjustments were made to ensure that all resources were effectively utilized and that no one was significantly overworked. Note the inclusion of an additional milestone (E-Proto) and the virtual elimination of major slope changes. The new plan now has the appearance of being "right" and is ready for use as the baseline plan when entering the execution phase.

In the revised plan, delivery of product to market was extended by about 15 weeks. After viewing this extension, the planning team from R&D studied the results produced and compared them with estimates generated during the definition phase. In so doing, the team found the new plan to be in far better alignment with definition phase estimates than the original plan. This fact, in combination with the detailed understanding of how the future was expected to unfold over time, led this team to think the revised plan was a much better forecast of reality than the first. It became clear that the original plan was of poor quality and that the revised plan was a significantly better predictor of future performance than the original. With R&D convinced it had developed a quality plan, it became possible to adjust the plans of other functions to accommodate the revised information. The new product development team and the organization had produced a plan of the execution phase that they felt could be realistically achieved. The organization was in the position to apply more resources to the effort to regain the lost 15 weeks if it chose to do so. In this case, in view of other constraints, they chose to accept the delay and entered the execution phase with the organization and project team working to the revised cross-functional plan.

In summary, the KMET chart can be used during the planning phase to get a rough idea (within a factor of 2) of how many people will be working on the project.

Second, slope changes can point to possible areas of concern within the network. A steep slope indicates that the running rate of task completion is about to increase. This is tolerable if it is accompanied by a corresponding increase in resources; otherwise, problems with task completion are likely to occur. It is especially dangerous to see a slope rise prior to a major milestone. Frequently, this signals a sign of wishful thinking. A reduction in slope from the normal running rate signals that forward progress has slowed down. At a slope of zero, the project is at a standstill! This situation can be eliminated by going back into the network to structure down the task links that are consuming excess time. A good plan should not return to a slope of zero. In this planning example, time to market moved from 61 weeks to 77 weeks. The 77-week plan gives the appearance of being highly achievable.

Analysis of the R&D branch of the hierarchical tree is now complete, but more work remains to be performed. To better understand the inner details of this analysis, a KMET chart for hardware, software, and systems needs to be produced. Likewise, marketing, manufacturing, and support must now perform KMET chart analysis on all other remaining branches of the tree given in Figure 4.6. In this way, the lowest-level network structures within the hierarchy will become fully visible to those doing the network analysis. To do a thorough analysis, a total of 17 KMET charts will be needed to fully summarize and explore the network detail contained within each branch.

As each branch is examined, attention is naturally drawn to the lowest-level branches on the tree. Higher-level summaries will point out the existence of potential problems; only detailed inspection at the lowest levels will isolate probable causes. Once probable cause has been determined within a given branch, the network will have to be revised until a workable solution is realized. Simple corrections can be performed directly on the tool. Complex and interactive modifications may necessitate redeployment of PLASTK within certain work groups to create a new network structure. Those involved with the initial planning will be called upon to review the analysis and help synthesize corrective network modifications.

Once satisfactory results have been obtained from the KMET chart analysis, plan analysis software (PAS) is used to generate a *scheduled completion report* for each branch of the hierarchical tree. This report is produced by sorting tasks within, or below, a branch by their scheduled completion date. The report provides a tabular listing of the detailed information behind each branch's KMET chart. The reviewer can examine the complete flow of work by task name, in the order it is scheduled to be completed over time. Information on the report is used to answer the question, Does the order of task completion make sense?

Review of the lower-level branches in the hierarchical tree is performed by those who generated the original plan, so as to check the results. Reviewers can look at the KMET chart for their respective tree branch as they scan down the task list. Things that don't seem right are easily spotted during the analysis. Work leading up to the completion of a milestone is examined in detail. In effect, the reviews are determining the quality of the branch's "to do" list. Once review and update is complete at the lowest levels of the hierarchy, the managers from R&D, marketing, manufacturing, and support can review the summation of tasks for their functions and make critical suggestions, corrections, and comments. Finally, the project/program manager reviews the entire listing for logical order, completeness, and accuracy.

All planners have now integrated the detailed task information with the information displayed on the KMET chart; the impact of the visual model (KMET chart) grows in significance to the planning team and the individuals that will work with the project's database. Now that plan network structure is correct, attention shifts to plan analysis with a focus on resources allocated to the tasks within the plan.

Most project management software contains certain functionality dedicated to resource analysis. Generally speaking, it becomes possible to view the work load of a resource over time. If person A is available to work eight hours a day, the software can analyze the work assigned to person A and display a histogram of hours worked above, or below, the eight-hour daily limit. Managers and those responsible for resource allocation at the project level can use this functionality to identify areas in which a resource is over- or undercommitted and bring this information to the attention of the individual and his or her manager. The person with the problem can then work to resolve the conflict while balancing other work responsibilities and update the database once a workable solution is produced. More often than not, the individual will have to interact directly with the project's database to determine what he or she needs to do to resolve the conflict.

For simple solutions to the allocation problem, resource leveling algorithms may be employed by the project/program manager at the top level. These algorithms adjust the sequence in which work is performed based on previously established project/program constraints. Sometimes the algorithms solve the problem. If so, the results are communicated throughout the hierarchy for review and acceptance.

Most times, the leveling algorithms fail to produce a complete solution. At this point, the impacted individuals, with support from a tool specialist, can interact with the project's database and develop a solution. These people and the appropriate managers can then review the solution and perform the final update on the database. At times, the solution has an effect on milestone timing, distribution of work among other resources, and the scheduling of other projects. If this is the case, cause-and-effect relationships need to be explored to the satisfaction of all affected participants. This interactive process will commonly be done on a concurrent basis by many individuals making adjustments to their assignments. If work on the front end was done responsibly, the changes produced by this dynamic process will usually have little impact on the overall schedule of the project; however, certain groups/individuals may have to adjust the distribution, timing, and allocation of their work.

Four plan analysis techniques have been described. Network analysis by phase and function summarization provided the first view of the project's overall timing. Utilization of the information gave the planning team the ability to make course adjustments by project phase and by functional work group. The KMET chart offered a way to visualize the structure of the networks making up the plan. Examination of the slope of the KMET chart identified areas for possible improvement. A scheduled completion report was generated to detail the information that had been graphically displayed in the KMET chart. Work groups, managers of work groups, managers of managers of work groups, and project/program management were given a complete "to do" list, sorted by time, for their areas of responsibility. Finally, the content of the project database was made available to individuals to give them the opportunity to make necessary adjustments. In all, plan analysis was done by a complete horizontal and vertical team that had a vested interest in the successful outcome of the project. Only a few steps remain before the planning phase is complete.

If your organization maintains a task work quantity database containing information on previous projects, now is the time to compare the new information to that in the database. Present task work quantity estimates within the project's database are compared with those from the past. Hopefully, you'll get close agreement between numbers if the historical information was effectively utilized by the modelers and planners up to this point. Critical questions must be asked if significant differences are observed. Most of the time, without the introduction of new strategies, tactics, or logistics, task durations will be a repeat of the past. If you are looking at a significant numerical change, make certain you understand the underlying reason(s). If the reason is based on fact and solid logic, go with it. If it is based on speculation and wishful thinking, go with history.

If your organization does not have a historical database, keep in mind that you are now taking the first steps in the production of this very useful item. Your present database will become history upon the execution of the project.

Lastly, before widespread distribution of the final project's targets, time is allocated to comparing the information content of the database with the estimates made in the definition phase. From the top level, how well do execution cost, complexity, full staff, and time-to-market planning targets compare with definition phase estimates? Refer to the matrix section of the Appendix and review each work group's targets against estimates. Keep in mind the convention that estimates are produced in the definition phase, and targets in the planning phase. Note that the "output" list within the matrix model has execution cost, complexity, full staff, and TTM figures in each cell. Where differences of 20 percent or more exist between corresponding numbers, detailed thought should be given to reconciliation of differences. It could mean a change in product definition, a change in plan, or both. But remember that this is the time to make the difficult decisions—before the commitment of resources, money, and time. Once closure is achieved, the planning phase targets are ready for distribution to management and the entire project team as the project's database is "frozen" to establish the project's baseline reference.

5.5 PROJECT TARGETS

At the top level, the final performance, development cost, and time-to-market targets are established for the project. They are summarized by each work group as specified in the matrix model contained in the Appendix. In support of these items is a complete project plan and corresponding database that detail the methodology by which the primary targets will be achieved.

Work groups (horizontal and vertical) throughout the organization shared in the development of the plan and are now positioned to begin product development. Project team performance during the execution phase will be measured against the targets established in the planning phase.

5.6   REPORTS AND CHARTS

The computer tools are used to generate five reports that are essential for communication of the plan to all team members and for communication of performance to plan and day to day status of the project. All five items are produced on a laser printer to accelerate the printing process. Descriptions of the items follow.

Individual "To Do" Lists.   Each team member receives an individualized report that lists all of his or her activities ordered by scheduled completion date. The printout is a computer-generated "to do" list for each person working on the project/program. Each task's duration and scheduled start date are also included. Space is allotted on the printout for actual start date and actual completion date. Utilization of these fields provides a convenient means for a person to report on progress against plan.

KMET Chart.   This is a graph of the total number of planned activities that are to be performed versus time as described in the previous section. Progress to date is plotted against the current plan; the original plan is plotted too, as a basis for comparison. The dates of critical events/milestones are highlighted. The program's completion rate, relative to the current plan, is also included. In the calculation of this percentage, only completed tasks are counted. This avoids the problem of estimating percentage completion for partially completed tasks. Tasks are counted only if they were scheduled to be completed, and were actually completed, thus avoiding the delusion that can occur if "easy" tasks are done first. A KMET chart is generated for each branch of the project's hierarchical tree so that visibility and detail are effectively summarized and reported.

Scheduled Completion Report.   This report sorts all activities by scheduled completion date, providing visibility of upcoming activities. A report is generated for each branch on the hierarchical tree. Each KMET chart is accompanied by its corresponding scheduled completion report to aid in interpretation and analysis. For each task on the report, scheduled start date, scheduled completion date, actual start date, actual completion date, and primary resource assignment to task are provided. Once the project is under way, this report becomes an excellent tool to chart progress that is taking place within each branch of the hierarchical tree.

Schedule Exception Report.   There are two parts to this report. The first lists activities that were scheduled to start but have not begun, while the second lists activities that were scheduled to be completed but are not finished. The team members responsible for each of these activities are listed too.

Alphabetical Task Name Report. This report is generated to help you remember what something is called. For those individuals that interact with the tools and the database, it is an indispensable item. As the database grows in size, the value of this report grows in direct exponential proportion.

Small to large projects can be managed to successful completion using the five reports listed above. Decision makers get the information needed to act in "real time" to keep the project moving. As the dynamics of the product development environment forces the plan to change, these fundamental reporting tools will help guide the change process and communicate results to the development team.

In some environments, it will be necessary to use the database and computer tools to generate other reports and charts. Frequently, project/program managers and managers of work groups supporting the project will be called upon to produce conventional PERT and Gantt charts. Without the availability of a very expensive electrostatic plotter, it is generally too time-consuming to produce output in this format when the database contains hundreds to thousands of elements. It's best to work within the organization to establish a report-generation system that eliminates the need for the production of plotted output; however, if this is not possible, the data exist to generate these charts. Similarly, as you get creative with the use of the information within the database, it is quite likely you'll want to produce specialized reports and charts unique to your situation. Again, the data exist to give you complete analysis, synthesis, and production freedom.

5.7 PLANNING SUMMARIZATION

Doing an effective job of defining, planning, and executing a project/program requires the existence of a life-cycle model at the foundation of the process. It should be evident that the model espoused in this book is divided into three distinct life-cycle phases (definition, planning, and execution), with clear focus on time to market. The execution phase was further subdivided into the prototype, pilot, and production segments. The output from each primary phase is specified using the new product development process matrix that is given in the Appendix. Chapter 4 shows how to structure the execution phase as a network while this chapter illustrates means by which network data can be analyzed to ensure a quality plan is in place for the execution phase to occur. From a modeling perspective, this structure may seem a bit unusual upon first contact and initial study. The primary reason for using this hybrid matrix and network structure is to provide a simple modeling framework that can be used to fit the project/program rather than attempt to model too much detail before quality information is available. A matrix modeling approach fits the needs of the definition phase and planning phase, whereas the network approach meets the needs of the execution phase.

From a commitment of resources and modeling perspective, this text advocates keeping the team size small during the definition and planning phases. Furthermore, the methodology stresses having the organization assign extremely talented people to perform tasks in the definition and planning phases. The combination of small teams and superior talent eliminates the need for a detailed network; however, the emphasis on output within the matrix is provided to maintain team focus on end objectives and to focus on measurable results. Too much is happening during a typical definition phase to even consider making a network of the activities and interactions taking place. Frankly, this is a situation that only the human mind can address. Key decisions will be made on limited data and will set the stage for the future. Work output from the definition phase drives the work done in the planning phase. During the planning phase, the execution phase network is synthesized. This network provides the entire team with the information necessary to complete the project/program.

Moving ideas to product reality is a real-time management and leadership challenge whose task and structural history used to be difficult to record. Since the execution phase of these projects was not structured as a network, it was not possible to record relevant data to help estimate and structure future projects/programs. Rather than provide network structure and tracking/measurement capability, many organizations collected "experiences" and recorded them in the form of "generic" life-cycle models. These old models suggest it should take so long to do a product definition and then about so long to complete development of the project. A record of detail that would substantiate the top-level information does not exist. Accordingly, the information is virtually useless.

It is now possible to maintain a history of what takes place on a project and make it available to those doing a different project in the future. Organizations with experience in successfully completing this transformation can build a database of knowledge that will be useful on the next project/program. With the new process and data collection equipment in place, the actual times needed to complete each task and the resultant task network connectivity will become historical information.

In this new form, the life-cycle model becomes a single time line. From a timescale perspective, the definition phase could extend over a few months to over a few years, depending on the complexities involved in getting technology and market into alignment. The planning phase can be done in a matter of a few weeks to a few months, depending on the complexity of the definition. The duration of the execution phase will be set by the plan. Clearly, the organization has some control over the amount of time devoted to each phase. The maximum that can be tolerated will be set by the marketplace. To simplify the planning of future projects and to make historical data easy to use, an organization can synthesize a "generic" network for the execution phase. In this case, the network would be created with tasks identified and durations synthesized from data on previous projects. Statistical analysis of task durations is of value when doing this exercise. If enough data exist, it will be possible to determine numerical values for worst-case, average, and best-case task completion times. Having a generic network available helps eliminate some of the difficulty associated with confronting a problem presented in this format for the first time. When a new project/program enters the planning phase, the project/program manager has his or her team of planners update the "generic" plan by adding resource information, new tasks, and revised linkages. Time-to-market calculations can be run on a statistical basis and the structure further evolved to reconcile differences between the definition and the plan. In this way, history is made available and reused in a convenient fashion.

The information presented makes no attempt to synthesize this "generic" network. Each organization must review its particular situation and produce the "generic" network that best fits its unique business needs. The "suggestions/requirements" section of the matrix model (planning phase) within the Appendix asks for a "generic" network model. Generally, existing life-cycle models can be transformed into network models in a matter of a few days. For those doing the transformation, the value gained in putting forth the time and effort becomes readily apparent. Scratch synthesis is also quite easy. To be effective, the individual teams (customer documentation, customer support, sales, marketing, quality assurance, manufacturing, personnel, finance, system engineering, software engineering, hardware engineering, project/program) should be asked to independently synthesize the network that relates directly to the creation of their output while working on defining the links with other teams. The combination of independent and interactive synthesis can lead to a quality network. This is also a good time to review the history of other projects to extract as much available task-timing and network data as possible.

A network would not be complete without progress reviews (milestones) inserted. This step should come last in the synthesis process to ensure that key milestones fit the project/program rather than having the project/program fit the milestones.

5.8   PROBLEMS SOLVED/MINIMIZED

Improved plan synthesis and plan analysis techniques can lead to the elimination of about one-third of the factors that contribute to project delay, increased development costs, missed performance goals, or a combination of the three. The 42 problem areas are listed in Section 1.3. Those that are impacted by good planning techniques are discussed below.

Inadequate Planning. Once a product development team has prepared a product definition complete with estimates, it may be tempting to begin the development effort without a planning phase. Frequently, managers and key advisors acting with the best of intentions will draft what they think is a reasonable top-down schedule and commit themselves and the development team to this rough, no-depth plan.

When this happens, it is the execution phase of the project that will be most out of control. Only top-level scheduling information will exist on the project. In effect, the project/program will have milestones specified over its execution phase, but the determination of milestone placement will have been performed by guess and by opportunistic thinking rather than by reason and logic. With a plan of this type, it is not possible to track and measure progress on a daily, or weekly, basis. Measurement of performance at a group or individual level is out of question. The only time measurement that can be made is at a milestone. Should milestone slip occur, and it is likely that it will, there is nowhere to turn to determine the cause of the problem other than to rely on the judgment of project participants.

Plans of this type undergo frequent revision at the top level. Of course, the direction of the revision is almost certain to extend the duration of the project and impact the product's time to market; however, most project/program managers find it quite easy to delay early milestones while keeping the product's introduction date fixed. Most people involved in new product development have some experience with this phenomenon. At the time of a milestone slip, you can convince yourself that circumstances were unusual and that it is bound to get better between now and the next milestone. Somehow you figure out how you are going to regain lost ground without spending more money or adding more resources. Besides creating a false sense of security for yourself, you'll probably come up with an argument so convincing that the management team of your organization will go along with it. Nothing new will really have happened until you approach the next milestone and it too begins to slip. At this point, you'll develop a new superficial plan. This time around, slip and delay become big issues. You'll have to move the product's introduction date, but to keep it from moving too far, money and staff will need to be added. With each tick of the milestone clock, the project grows in staffing and spends more money each day for more days as the product's introduction is delayed. Eventually, the project gets done. Sound familiar?

Organizations that do inadequate planning and find themselves in a new product development somewhat like the one just described must seriously consider changing their mode of operation. Most of the tools and techniques needed to make the transition are described in this and the previous chapter.

Fundamentally, each leading member of the new product development team should be actively sought out for input on his or her portion of the project. People can model their plan using PLASTK while developing and refining their technique. The information collection process should extend horizontally and vertically throughout the organization. Wherever possible, team members should be encouraged to make their contribution to the overall planning effort. Doing so will build confidence in the resulting plan, as well as give the project manager the opportunity to work with team members on a one-on-one basis. Sharing the planning load tends to produce a plan that has commitment as its foundation!

Frequently, as you begin the shared planning exercise, you will observe individuals who have moderate to extreme difficulties thinking in terms of tasks, task durations, linkages, and networks. This occurrence generally requires that you be able to coach the people having the problems and provide on-the-job training in doing planning compatible with simple tools. If training assistance is available, it is wise to do as much as possible before asking for inputs. You should not be surprised that the ability to produce good plans is not a universal gift. If coaching and training fail, the planning task must be performed for these team members by someone with the necessary skills and talent. When entering the execution phase, keeping a close eye on those who had difficulty with planning tasks is advisable. They usually tend to have similar difficulty in staying on top of the work they need to accomplish.

Excessive Planning. While doing too little planning is somewhat of a problem, doing too much is often the error made by many new product development teams and their managers. Planning and replanning and planning and . . . This cycle never lets the planning phase end and the execution phase begin. It may indicate that the planning process needs improvement or it may forecast serious problems with the stability of the definition that is driving the plan. In either case, an understanding of the problem must be reached and the cause eliminated.

It's usually easy to determine if a changing definition is at the heart of the problem by asking those involved in the continuous planning process for their feedback. In addition to this action, review of the external reference specification (ERS) revision log will help develop added insight. The mere fact that the planning phase can expose such a serious, fundamental problem is reason enough to take the planning activity very seriously. Clearly, the only right action to take is one that will stabilize the definition to allow for successful completion of the planning phase.

Problems with the planning process itself comprise the other major contributing factor to this problem. Most of the time, the problems can be grouped into technique and tools categories.

From the perspective of technique, the most common error is that of trying to accomplish all planning activities through a "super planner." This individual tries to do it all! No shared and distributed planning techniques to worry about here—all plans get done by one person, and of course there are resultant consensus problems. A second pitfall to avoid is a project master plan with excessive detail. To prevent this, it is necessary to have individuals, while managing their own work affairs, provide summary task information at the project level. Where possible, individuals should be encouraged to generate task information that is two to four weeks in duration for tracking at the project level. It is assumed that individuals will maintain detail with greater resolution at the local level.

Problems with tools are usually associated with software, hardware, and utilization techniques that are too slow. If the speed of tool performance and utilization is slower than the planning or plan revision process, then the tool serves no useful purpose. Following the selection procedure at the beginning of this chapter will ensure the wise choice of tools and peripherals.

Doing a Job for the First Time. Most project/program managers are new to their assignments, as will be many managers and contributors across the matrix. In fact, it is typical for over 50 percent of a new product development team to be new to their assignment. Thus, these individuals need to learn pertinent skills as rapidly as possible. Getting the project defined is a major step forward. Getting the project accurately planned will be the next major learning opportunity. The synthesis of networks, task estimation, resource allocation, both cost and time predictions, and dealing with bottom-up and top-down pressures will all be new challenges that must be met by members of the new product development team.

Fortunately, many organizations have recognized the need to improve the planning skills of their new product development team participants. Training programs (internal and external) are produced and conducted to develop a reasonable skill set. In most cases, training will focus on project management mechanics (networks, PERT, Gantt, etc.) and financial tracking methods. Frequently overlooked are the training needs in the areas of team building, interpersonal relationships, shared planning techniques, tool selection and utilization, and global time-to-market considerations. The thought of improving the planning skills of those not involved in project/program management is a foreign concept in most organizations.

Those new to the organization and many individuals with years of experience may find themselves struggling with planning concepts. Since formal training in hierarchical network planning is not likely to be a part of the on-the-job experience, the project/program manager will often be relied upon to guide others through the planning process. If this manager's skills are not the best, the entire planning of the development effort may contain critical flaws. Therefore, the planning process that is followed should contain a number of checks and balances to ensure the production of a realistic plan under less than ideal conditions.

The production of a quality plan by the entire development team provides an immediate solution to the problem of doing something new for the first time. With product definition in hand, the team goes through the steps outlined in the planning process flow diagram shown in Figure 4.1. Project/program managers lead and facilitate the planning process, but are removed from creating the actual plans for work groups and individuals. The entire horizontal and vertical team works together using PLASTK as a plan synthesis tool to figure out how to accomplish the development of the product. The team is asked to use skills it already possesses to take this first step; it is not necessary for individuals to be taught new skills. Effectively, the planning process removes the "newness" from the development process. After all, once serious thought is given to a particular problem, the uncertainty in the development of a solution has been reduced. Shared planning brings the development team together and gives the participants the opportunity to learn about and then create the structure that will guide their future work.

To ensure that the plan is realistic, the plan is input and analyzed by a computation system. Feedback, from five simple reports and charts, is generated and supplied to the team of planners to give them the opportunity to make adjustments to their original structure and data. The close coupling between synthesis and analysis serves to develop planning skills, without the need for formal training. As skills improve, so does the quality of the resulting plan. Within a few iterations, the output from the planning process will converge on the production of targets that are fully understood and committed to by the product development team. With the production of the targets comes the maturation of the work force needed to begin the execution phase effectively.

Overwork/Burnout. The concepts of the minimal serial network and determination of project timing by the application of resources to tasks make the overworking of resources very visible. Once data is in the computer system, viewing how each resource works over time is easy. A resource histogram for each person is displayed and analyzed. Where resource are over- or underutilized, adjustment in their planned commitment to work on the tasks in question is required. The individuals allocating resources to tasks over time maintain control of the pace at which the project will proceed by the way they structure, estimate, staff, allocate, and run the project. With the minimal serial network in place, product can be delivered to market earlier simply by carefully applying resources to reduce the time needed to complete critical path items within the network. The decisions made in planning will set the early tone for the rate of work progress. A good plan will provide a reasonable distribution of work across the entire team, with everyone contributing their fair share.

The creation of a project/program network can be the best action to take to minimize the occurrence of overwork/burnout. Project teams that know what is ahead, what has been accomplished, where to deliver their work, and where the big problems are will have the focus necessary to direct their energy into the critical few items necessary to project success at a particular point in time.

The planning team can take data from the project management software used by the project and sort the information by resource and scheduled task finish date—this results in the production of the computer-generated "to do" list. With this list, the work load each individual carries becomes reasonably clear to the individual and his or her manager. Data sorted in this fashion can help balance assignments by identifying those who are underworked and those who are overworked. Also, the planners should get involved with those responsible for the tasks to work out task finish dates that are realistic. In this way, product development contributors and their managers can reach agreement on what is to be done by when. Individuals who are overworked in the plan are given the opportunity to correct the situation by adjustment of their "to do" list before the execution phase begins.

Additional Time Needed. Most of the time, task duration estimates tend to be optimistic. Little attention is given to historical data. People providing the estimates tend to respond with how long they expect something to take and not with their estimate of the work-hour content of each job assignment. It is most important to stop the practice of asking "How long?" and ask instead, "How much?" Making this conceptual shift in doing estimation when developing a new product is the major factor in the production of quality task duration targets. Estimators should never be asked how long they expect a task to take. Instead, ask them to provide estimates of work content and suggest reasonable work rates, which results in the computation of "how long." The PLASTK tag in Figure 4.4 initiates the thought process. This focus makes it possible to rely on information from data on similar tasks done in the past. Previous estimation errors are corrected, and people begin to improve their estimation skill techniques. There is now a strong tendency to base estimates on things going right and some things going wrong; in effect, the estimates are padded for unforeseen circumstances. Inherently, if the task could be worked on full-time, it would get done in a minimal amount of time. Sometimes extraneous factors will prevent dedicated work on the task, and more calendar time will be needed to finish it. Under these conditions, the additional time needed is not caused by poor estimation; it is the direct result of not working on the task. Making task duration extensions visible during the execution phase enables easy control of them.

Members of the new product development team making the estimates must guard against this common error of inaccurately padding estimates. A good check is to compare new task time work quantity estimates with those from previous projects.

In general, this information is readily available if the methods described in this text have been followed, but it is often difficult to extract relevant data from other new product development methods. Think of how difficult it is to compare how long it took on project X to do a task with how long you expect it to take on the new project. Basically, comparison of reduced durations is impossible. Saving the data offers little benefit for new product development process application and improvement. On the other hand, work quantity data is extremely valuable. If recorded work quantity data do not yet exist, another control to improve estimation accuracy is to take the new estimates and ask an experienced person to review them (people with experience tend to be better estimators). Where differences are detected, it is good practice to have the individuals involved reach agreement so that both improve their estimation skills. One additional method is to develop time estimation models, such as those given in Figures 3.3. through 3.6, for major tasks frequently performed by the new product development team. These models can be used to get plans and models to converge on one another. Divergences should be noted and analyzed to ensure correctness.

The goal is to become skilled in task estimation so that the need for additional time to complete a task becomes a computed parameter that depends on the amount of effort expended to complete a given task. Tasks that are not on the critical path can be allowed to take longer provided that resources needed to complete these tasks are first directed to complete work on critical path tasks first. Control of project execution time becomes a matter of application of resources to the right tasks at the right time.

Long Turnaround Time. The more parallel the structure of the new product development network, the higher the likelihood of turnaround time delays when work enters a department of work area with finite capacity. For instance, a product that consists of 10 printed circuit assemblies faces a serious challenge if all designs are given on the same day to a design department that can only accommodate two new board designs per month. In this case, two of the 10 boards would exit within a month, while the remaining boards would exit over a period of four months. Similar challenges are encountered when utilizing scarce, expensive design systems, newly developed integrated circuit fabrication processes, advanced software development tools, and other capacity-limiting factors. One of the serious challenges confronted by concurrent engineering teams is when a number of parallel efforts are forced to go through a highly restrictive serial constraint. If money is available to grow capacity, by all means do it. However, if this is not an option, then input to the asset must be controlled to extract maximum financial benefit.

Utilization of areas with finite capacity, like all other resources, are placed within the project's network model. Printed circuit design systems, integrated-circuit design systems, fabrication equipment, and other key nonhuman resources become an integral part of the organization's resource pool. Let's use the 10-board example again. A task like "design board #1" is thus staffed with the person that will do the work and assigned the tools that will be utilized, as the remaining nine such tasks would be; the people and tools assigned come from the resource pool available to the project. The computer system can then be utilized to determine the optimum sequence of work through the resource-limited area. This is done by intelligently adding parallelism around the serial flow through the constraining area. Things can be done to establish the right order of input to the printed circuit area. Flow work in as it gets done. Where possible, save the low-risk work around boards for last. As designs are produced, structure the assembly of the product in a progressive fashion that is paced by factors other than the flow of printed circuit boards. Take small steps on large jobs. Consider the percentage of boards that will require revision; where possible, slow down and get the job done right the first time! Look at the time that can be saved if this percentage is cut in half. Collect all thoughts and represent them with a network model. Adjust the model until a degree of satisfaction is realized. Surprisingly, once networks get structured around the serial path, the critical path often becomes a part of the peripheral network. Ideally, adjustments should be made to the peripheral parallel network until it is equal in length to the serial network.

With the network structured, and resources assigned, the computer tool is used to observe a resource histogram of the areas with limited capacity. The histogram will reveal where capacity is exceeded and where light utilization periods are situated so steps can be taken to better balance the work load. A few iterations between commitment of resources to task and task network structure are all that may be required to make optimal use of a very limited asset. Done in this manner, contributors to the asset know when their input is expected while those doing the work receive some indication of the work input they can expect to receive. Both sets of participants become aware of the planned sequence of events and become sensitive to each other's needs.

Frequently, the capacity issues can be adequately resolved if support groups are given reasonable advance notice to prepare for the "crunch." An accurate plan provides the notification well ahead of the actual need for service. When an entire organization runs in this fashion, that is, using realistic, accurate, well-communicated, and trusted plans, it is rare for turnaround times to go beyond their normal windows. A fringe benefit of predictive accuracy is that it provides an incentive to tackle critical demands when they do arise. If they are the exception rather than the rule, there will be an invigorated sense of urgency to "put out the fire" and really get moving on a key defined project/program needed.

Internal Competition. A major factor in delayed market entry of new products is internal competition for resources. The competition is almost always a result of an organization staffing many projects in a constrained manner rather than fully staffing a critical few projects needed for business success. It's hard for some companies to say no to some ideas and pick only the best available alternatives. Running too many projects, each staffed well below full staff needs, forces all projects operating in this manner to have delayed market entry. An organization doing only "first-of-a-kind" products may be able to operate in this manner if it is years ahead of the competition with its technology and product implementation ideas. However, most other efforts will have distinct windows of opportunity that must be met to avoid severely degraded business performance. In business ventures where market-entry timing is of high priority, rivalry between projects for scarce resources must be brought under control so that each project receives only the resources needed at the time when they are needed. Sequencing new product introductions and fully planning the sequenced application of resources to make them a reality become critical functions of the planning process.

Allocation of resources to the project, as described in the resource allocation branch of Figure 4.1, commences the exercise in the determination of how many new product development efforts an organization can realistically support. Development of an organization's resource list places finite limits on the resources available for allocation to projects. Just creating this list elevates the awareness of exactly how many resources are dedicated exclusively to new product development. New, knowledge from the business definition process and information available on new products presently under development are summarized and placed alongside the resource list—this is the list of projects available to the resource allocation team. All that needs to be done is determine the sequence of how to move products to market and then assign resources from the resource list over time to make product delivery a reality. This is no simple task. This is why a tool such as the resource allocator (RA) was developed. With RA, managers with the authority to assign resources to projects do so as they determine the initial sequence of product introductions for the organization's new product development program. This action forces members of top management to begin considering just which projects they should undertake as they establish preliminary resource allocations to each project based on requests for resources developed by project participants during the definition phase. RA makes the allocation of resources to projects highly visible. In response, resource allocators begin the process of reducing the number of projects identified for full staffing.

To further trim the project list to the number that can be fully staffed, an organization must merge project database information of those projects presently in the execution and definition phases with those about to emerge from the planning phase. Taking this action activates using the project management software's functionality to merger projects and subprojects discussed in the beginning of this chapter. Merging of multiple projects allows high-level resource allocation managers to view resource utilization over multiple projects as a function of time. In so doing, they can assess when each project is expected to come to market and thoroughly review the commitment of resources to each project. Performing an integration of multiple projects as well as a thorough analysis of each project's database makes it possible to staff for effective execution by limiting the number of projects in the execution phase and by fully staffing and supporting those that are in full development.

No longer will new product development teams be treated on a first-come, first-served basis. Prioritization, selection, staffing, and sequencing become essential in effectively utilizing limited resources. Internal competition must be minimized so that energy can go into getting the required new product work done to enable the team and the organization to compete with external competition.

A vital, time-to-market oriented organization will encourage competition through the definition and selection of projects/programs, but will then put emphasis on establishing a plan that will get the product to market in the most efficient manner possible. Good planning builds the organizational structures necessary to realize these objectives. Every project/program under way in the organization needs to be well planned to ensure that adequate capacity is available to achieve timing goals and to minimize unproductive behavior.

Management-Driven Schedules. A management-driven schedule is one where top management as a team, or sometimes a key management figure, establishes the milestone sequence for the new product development effort. Management dictates the effort's start date, what will be done, where the resources will come from, and the effort's expected completion date. Generally, few details about project execution are supplied. Being on the receiving end of such a directive can sometimes seem rather oppressive. However, there may be a positive side to this situation. First, if the management team is competent, it has taken a major step forward in firmly establishing the product's definition. Only a management team that knows what it wants would ever dare take this leadership position; it is unlikely that it would then abandon its vision or want to see it compromised. Second, because of the high-level attention received by the project, it is likely to be fully resource-funded. Third, like it or not, the timing goals of the project are well established. Making the schedule visible may have a big effect on cross-functional commitment to achievement as top management elevates the "can do" spirit of the organization. A challenge coupled with high uncertainty can sometimes be the catalyst to do great things. Before going forward with unbounded enthusiasm, however, a few individuals on the new product team must reflect on the situation and plan its future. A management-driven schedule supported by an aggressive but comprehensive plan is likely to succeed, but one that is not cross-functionally planned and orchestrated will fail.

Think of a management-driven project as a normal project that has its definition phase significantly reduced. Of course, no project or management team after reading this text would want to go forward on project execution without conducting a brief, but intense, planning phase. As previously discussed, one key reason for doing the planning phase is to measure the stability of the product's definition. If the definition's content is changing so fast that the planning team cannot keep up, probably the definition is not yet stable enough for the project to enter the execution phase. On the other hand, if a working plan can be produced, then the management-driven project passes its first critical test and receives a vote of confidence. Results from the planning phase can be used to establish a dialog between the project and the management team. Plan synthesis followed by plan analysis, as described in this and the previous chapter, leads to the production of data that can be used by both parties to produce a win–win situation. Adjustments may have to be made based on the data assembled by the planning team, but once agreement has been achieved, the new product development effort is fully supported vertically and horizontally throughout the organization.

Plan synthesis and plan analysis techniques effectively applied in management-driven projects can really help an organization deliver product to market in a timely fashion. The new product development team that can show meaningful, well-reasoned relationships between the variables will get the support of management as key problem areas are identified, discussed, and resolved. A project team that does not conduct a planning phase is not only gambling with management support, but is also risking failure.

Inadequate Progress Reviews. It is during the planning phase that progress reviews will be specified for the execution phase. During the latter, the milestones that were selected are likely to be closely watched by top management and all participants of the new product development effort. Observation of major milestones and events is certain to lead to measurement of their occurrence as a function of time. Through milestone measurement, project progress will be observed as opinions about the overall state of project affairs are formed. To keep the horizontal project team and vertical management team working as a unified force requires the identification of quality milestones spaced at appropriate time intervals. The milestones must be significant to project and management new product development participants and at the same time exist in optimal quantity. Too many or too few can be detrimental to top-level summarization of project information.

From a quality perspective, the purpose of milestones is to encourage the production of decisions that lead to action by new product development participants. Key information needed to produce the decisions must be produced and effectively displayed to facilitate key project decisions. In any new product development effort, three milestones are dominant: completion of the definition phase, completion of the planning phase, and completion of the execution phase. This text provides detailed information defining when each of the three fundamental project milestones has been achieved. Suggestions have been provided on how to structure and summarize information to enable the production of accurate and timely decisions. The KMET charts, resource-leveling summaries, and collection of "to do" lists produced at the end of the planning phase are certain to aid the decision-making process. Production of these summary items relied on tools such as RA, PLASTK, and PAS. Utilization of each tool helped to isolate variables that were then used to produce key project decisions. Planning decisions progressed from allocation of resources to the project by management to the actual production of a detailed implementation plan for the project in a logical and intuitive fashion. With each step, visual evidence of performance was supplied so that participants could rapidly and visually evaluate their work under dynamic conditions.

Consideration must be given to the identification and specification of milestones that are needed between the three primary milestones to ensure that an adequate number are in existence. For an organization following the recommended staffing profile documented in Figure 1.1, consideration should be given to having time-driven milestones during the definition phase, no milestones during the planning phase, and event-driven milestones during the execution phase. Initial projections of how long the project is expected to take should be used to establish definition phase review points at the 25-percent, 50-percent, and 75-percent intermediate points of the definition phase. Of course, the final definition phase review occurs at the conclusion of the phase. The scheduling of definition phase reviews is done using the following equations:

$$\text{Review 1: R1} = \frac{r}{8}$$

$$\text{Review 2: R2} = \frac{r}{4}$$

$$\text{Review 3: R3} = \frac{3r}{8}$$

$$\text{Review 4: R4} = \frac{r}{2}$$

where r = expected duration of the project from the start of the definition phase to the end of the execution phase (refer to Figure 1.4 for a graphical description). For example, a new product development effort that is expected to require two years from start to finish should have definition phase milestones scheduled at three-month intervals where R1 = 3, R2 = 6, R3 = 9, and R4 = 12. Projects with a shorter definition phase will have the same number of reviews except they will be held more frequently. The time between reviews extends for first-of-a-kind and next-generation efforts operating within a longer time-to-market window.

Because the planning phase must be short and intense, no intermediate milestones are recommended. Measurement of the beginning and the end of the planning effort is all that is required. Four weeks is nominal organizational performance. If this phase is longer than six weeks, the cause for the delay should be investigated.

Execution phase event-driven milestones should be defined by the plan and be tied to the production of key output items as the execution phase unfolds. Suggested examples in Figures 5.5 and 5.6 include milestones called "Design Review," "End-Proto," "End-Pilot," and "End-Production." During the design review, members of the new product development team check each other's work to ensure that all aspects of the product's definition have been effectively addressed by the design. At the end of the design review, sufficient information exists about the product to begin its actual realization. At the end of the prototype phase, the actual product will be functioning and team participants are able to measure its performance and determine cross-functional impact and action. A pilot unit is well on its way to being ready for delivery to customers. These units will be of final configuration but may be built using nonstandard processes, with help from nonmanufacturing personnel to work out the "bugs." The "End-Production" milestone is achieved when the execution phase is complete, and that is when the product is ready for customer delivery and all key participants within the organization agree that this is true.

Determination of exact execution phase milestone nomenclature and requirements at each of the milestones is an exercise every new product development organization must undertake. The starting point is the time axis of the organization's matrix model. In this case, as was the case with the functional axis of the matrix, the primary timing information of Begin Definition, End Definition, Begin Plan,

End Plan, Begin Execution, and End Execution is hierarchically decomposed to another level with a bit more content and time resolution. Methods used to establish the milestone sequence for the organization and the project were discussed in Chapter 4. Once determined, the sequence of milestones becomes part of the planning methodology used by the entire organization.

From a quantity perspective, the number of milestones specified over the complete duration of a project should be held to approximately 10: 4 for the definition phase, 2 for the planning phase, and 4 for the execution phase. Creating more than 10 is not necessary when consideration is given to the fact that a full matrix model and network model exist for all projects. Using matrix and network new product development process models eliminates the need to track a multitude of top-level milestones to exert some control over a project. Remember, under the conditions described in this text, a task network is resident between milestones. The performance of each task to plan can be observed and measured. Project management attention is focused at the task level. Milestones are merely summaries that can be used by others interested in observing the performance of the overall project as it moves from one milestone to the next. Where shifts in milestones are detected, attention moves to the network model to determine cause-and-effect relationships.

Testing Turnaround Time. Just how much work goes into conducting a test? I was able to observe an associate of mine, Dave Gildea, collect data on a project he was managing that would help me produce an answer. My thoughts on this topic were not unique—they were inspired by developments in software engineering that could be used to predict the number of defects per 1000 lines of code and the number of test hours required to find the defects. What I did do was to take software theory and apply it to unified hardware, software, and system new product development. Additionally, I took advantage of the data Dave collected and the timing of his project with respect to mine. I knew the hardware (number of components), software (number of lines), and system (number of modules) of Dave's new product development effort, which was in its execution phase. To assist me in answering the question of just how much work is involved in conducting a test, Dave asked his new product development team to maintain a log of problems discovered with the hardware, software, and system. Each problem was counted, classified, and tracked to its eventual resolution. While Dave collected the data, I was able to use it to perform some simple calculations to arrive at the answer to my question.

I was now in possession of data on the number of defects that my project could expect to have, per 1000 hardware components and 1000 lines of software contained within each module of the system, during its execution phase. Moreover, I had information available that helped me estimate the number of test hours needed to find each defect. In effect, I had all the information needed to help me make a sound estimate of the number of test hours my project would require based on its complexity. I then used this information to produce a network model of my project's test-related tasks; the model was therefore based on data and fact rather than on gut feel or an overoptimistic guess.

Information of this type makes it possible to structure and staff test-related tasks effectively during the planning phase. It is a way in which future problems can be anticipated and addressed, even when you don't know what the problems will be! It is a key benefit derived from having a task and work-hour database. Maybe the first project that generates the information won't benefit, but future projects certainly will. On the one hand, no one really wants to plan for something going wrong; on the other hand, planners must strive to accurately forecast the time to test the product and correct deficiencies that have been observed. Fortunately, accumulation of this data now makes it possible to accurately predict test time as a function of product complexity. With each data point collected, statistical accuracy of the database improves.

Development teams and their managers must make the collection and maintenance of new product development process data a high priority to extract maximum benefit for the planning of future projects.

Management Methodologies. The backbone of this text is management methodology applied to the development of new products. Applying the suggestions makes it possible to improve an organization's ability to output new products. Nowhere in the overall new product development system described in this book are management methods made more visible than in the planning phase. This three- to six-week period of time measures how well the definition phase was executed and prepares the new product development organization for what it is to accomplish in the execution phase. In effect, the planning phase serves to measure the management methods that lead to the production of the plan itself. An organization that can synthesize a complex cross-functional plan, analyze the plan, and then use the data within the plan to make key new product development decisions is likely one that is effectively managed.

The production of high-quality plans for each new product development effort suggests that management knows how to conduct a definition phase. If business definition issues and product definition topics are changing at a rate that is faster than that at which the organization can produce a meaningful new product development plan, the new product effort will surely experience introduction delays, cost overruns, and performance deficiencies. This results because cross-functional product development activities are not coordinated and specified by a reasonably stable plan—working with some structure and guidance to instill some discipline in the new product development process is absolutely essential to process success. Plans become the means by which organizations are linked both horizontally and vertically as a steady flow of new products to the market is produced as a function of time. They produce the data necessary to create an accurate visual illustration of how an organization can realize its business and product visions over time based on realistic application of its limited resources to the new product efforts it elects to staff fully.

Successfully managed companies make their organization's ability to produce accurate plans a high priority. Project teams use plans to "ask" management for the resources needed to execute their projects. Managers merge the requests into current operations to determine where and when resources are likely to become available.

Furthermore, as plans are being used to study resource allocation alternatives to each project, strategic business unit and functional managers are deciding the priority of projects they would like to have enter the execution phase. An organization's ability to produce timely and accurate plans at the project level relates to the overall business effectiveness of the company. As individual plans are integrated and merged, key decisions are produced on the number of projects that will be taking place and the commitment of resources to each of the projects. Completion of the merger and analysis of plans results in management "granting" resources to selected requests. By taking this action, a company elects to fully staff only the "best" projects that emerge from the planning phase. "Best" is a term that is fully understood in the same way throughout the company—the result of having a unified business definition. Decisions will also be made to terminate work on less desirable projects—planning has a way of highlighting the undesirable properties of certain projects. In this manner, project teams and their managers work together to determine which projects move forward and the effectiveness of the resource commitment made by the organization to the selected projects.

People's Skills Do Not Match Job. In the planning phase, resources are allocated to projects by a management team, which does not possess detailed task knowledge, using the resource allocator (RA) tool. Managers making these allocation decisions can sometimes make an error. Errors are detected when resources are assigned to tasks during the modeling of the new product effort or later, when actual resource assignments to task are made within the computer system. When mismatch between task resource need and resources available to the project is encountered, resource allocators at the project level need to discuss the situation with managers making resource assignments to the project. In the spirit of cooperation and with detailed project task data available, the team of managers can collectively work to develop a solution. Following the techniques described in this chapter and the previous one will almost certainly prevent the "mismatch" problem.

A good plan will match skills to the task or it will identify the training requirements needed to get people prepared to do the job. Where training is needed, this task can be called out and incorporated into the project/program network to ensure its impact is factored into the overall structure.

5.9   GRAVIS ARBITRIUM

With planning activities complete, it is now time to produce *gravis arbitrium*—the important decision.

From a process perspective, the business definition, product definition, and execution phase plans are complete. If the staffing and development cost profiles discussed in Figure 1.1 have been followed, half of the time required to get the new product to market and 20 percent of the funds needed to develop the idea fully have been consumed. Under these conditions, enough information has been produced by the new product development team and its immediate managers to request the production of a top management decision—one fully committing the organization to the plan or terminating the entire product development effort.

As the new product development process matrix (see Appendix) indicates, upper management must feel comfortable with

1. *The Stability of the Product's Definition.*   Has the plan been produced in a short, intense period of time? Does the planning team seem confident in the results of its work or is it constantly modifying the project's database? Are summaries in place that look reasonable? Are business objectives being met?

2. *Matrix and Network Models of the Execution Phase.*   Are the networks minimally serial? Have the majority of tasks been identified?

3. *Resource Allocation to Task and Time Duration Computation of the Execution Phase.*   Have all task durations been computed by the application of resources to tasks at realistic rates? What do the resource histograms of each resource assigned to the project look like? When asked if results can be achieved earlier, do managers of the effort immediately shift their attention to areas in the plan where they know they are resource-limited? Can time to market of the project be modulated by the application of resources to the model?

4. *The Appearance of the KMET Chart.*   Is it linear? Can slope changes be effectively explained? Based on what is delivered, can support be provided to make minor adjustments by the top management team?

5. *Closure of Definition Phase Estimates With Planning Phase Targets.*   Do the results produced by bottom-up modeling and targeting closely agree with definition phase estimates across the NPDP matrix? Does the product get to the market in time? Are development costs under control? Have productivity figures been effectively utilized in the targeting and estimating process or are numerical values in the plans database purely speculative?

6. *Integration With Other Projects Sharing the Same Resources.*   Have resources been leveled across the concurrent projects to which they have been applied? Are all projects fully staffed? Is it clear which effort would be terminated if the project under question were to meet a serious challenge requiring additional resources?

A management decision to terminate the effort keeps expenditures well in control as attention shifts to redeployment of resources to other projects.

A management decision to proceed activates the plan. Over a short period of time, staffing grows from 40 percent of full staff needs to 100 percent as the project enters the execution phase. This phase is fully explored in the next chapter.

6

EXECUTION

In this chapter, you will learn about

Tracking a project on a real-time basis.

The automatic generation of tracking reports and graphs.

The synthesis of execution strategy, tactics, and logistics on a real-time, dynamic basis.

Modification of the project task network.

Problems that can be solved or minimized by the application of good execution techniques.

Use of the KMET chart in executing a project.

Recommended actions following the execution phase.

The purpose of a team is not goal attainment but goal alignment.
                                                        Tom DeMarco and Timothy Lister

6.1 TRANSITION INTO THE EXECUTION PHASE

Synthesis is often not associated with the act of carrying an activity to completion. The game of soccer serves as an example of execution synthesis. Rules are set and the game is well defined. Prior to kickoff, a game plan has been put into place. With each tick of the clock, the game is played out to determine the eventual winner. Each player on the field works to synthesize the next offensive, or defensive, play opportunity. Within the limits of the game, all participants strive to maximize their performance by setting up and executing their next moves to the best of their abilities. The players' minds are at work. The action stops only when time runs out. *The synthesis process is dynamic and is carried out by all persons on the field with leadership provided by key team members and the coaching staff.*

Similarly, the execution of a project plan is a dynamic process that frequently requires synthesis to take place in a very short time frame relative to the other time frames in the definition and planning phases. Definition decisions tend to have the luxury of time. There is pressure, but the proper synthesis of a decision generally takes months, and sometimes years. Upon entering the planning phase, the process speeds up. Plans can usually be synthesized in three to six weeks. During the execution phase, decisions need to be made in days or hours. Time becomes precious. The penalties for letting it slip by are severe. The transition into the execution phase must be managed to ensure that the new product development team has an effective start.

For process participants, a new perspective on the product development effort must rapidly develop. Attention shifts from thinking, exploration, and planning to action. Now, team participants have to actually produce the abstractions contained within the product's definition. For some new product developers, this is a difficult period. Engineers and technologists sometimes want to continue refinement of the technology that will be used to create the product; marketers start to have second thoughts about feature sets and performance numbers; and manufacturing may have doubts about how certain critical processes will perform. Also, leadership and management styles must undergo a rapid revision. An informal and sometimes casual control system must be replaced by a system that runs more by numbers and data than by intuition and feel. Too much is happening too quickly on a concurrent basis to comprehend the situation without process measurement and data analysis. So, as team members are making the transition to the production of real output, they also become aware of the fact that what they do will be measured on a periodic basis. In effect, the entire team moves from preparing to be in the execution phase to actually being in the execution phase! The reference plan becomes the competition. If the team can perform to or slightly ahead of plan, a winning situation is in progress. When progress departs from plan, participants must be prepared to make adjustments to recover and retake a lead position with respect to the reference plan. At no time does the process come to rest until end objectives have been achieved.

Many new product development team participants know that the execution phase of a project takes place a day at a time. Each day is extremely important and a day wasted working in the wrong direction cannot be made up later. The thorough, well-structured plan produced in the planning phase makes it possible to track the execution of the project on a real-time basis so that trends in work progress can be monitored and measured as a function of time. In almost all cases, if the managers of the plan are good synthesizers of tactical and logistical maneuvers, problem areas relating strictly to execution can be detected early enough to take corrective action, thus avoiding a slipped project milestone.

Since early warning and immediate action are the key ingredients in controlling execution problems, a task tracking system that provides quality information is essential. The computer tools selected in Chapter 5 are used to perform this function. As time goes by, the date of commencing work on a task is recorded and compared to the start date of the task specified in the plan. Deviations from planned start are observed by the management team. Questions will likely be asked in regards to those areas that appear to be off to a slow start. The plan can serve as a point of comparison and help direct attention to potential problems after the execution phase is only a few days old. Once a week or two has gone by, data will become available on whether or not planned tasks have actually been executed. Such data on actual project performance as compared to plan are utilized to help the project team make an effective transition into the execution phase.

Figure 6.1 illustrates how to guide the new product development effort into the execution phase using only a limited amount of data. In the figure, two activities are

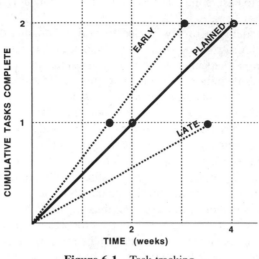

Figure 6.1   Task tracking.

planned to be performed by a resource in the first month of the execution phase. The first activity is scheduled to be completed in week two, and the second in week four. Thus, the "planned" line shows the accumulation of two tasks over the first month for this particular resource. To simplify measurement and to avoid any ambiguity, tasks are considered complete when they are complete. No credit is given for partial accomplishment—the task has to be 100 percent done to be recorded as complete or the model must be adjusted to reflect the new situation. This can be done because the resolution of computed task durations contained within the database has been structured so that each resource is scheduled to complete one to two tasks per month. Data points spaced at two-to four-week intervals are all that is required to make informed interpretations of trends contained within the data. At two weeks into the execution phase, a trend line for this resource appears. If the first task was completed as planned, it will probably draw little management attention. However, if at the two-week point the completion of the first task is early or late, determination of preliminary cause is justified. Being early could mean that work quantity was underestimated or that the resource commitment was larger than necessary. A late task completion may be caused by a low work quantity estimate, too small a work rate on the task, low productivity, or a combination thereof. So, just two weeks into the execution phase, data has been effective at directing management attention to areas where potential problems may exist.

I've made the transition into the execution phase on a number of projects, but only on the last two major programs were plans and comparative data utilized for rapid adjustment to the execution phase. We called the individuals that managed at the second level of the hierarchical tree "Green Guys" because their branches appeared in green on the hierarchy chart. In Figure 4.6, the program manager and the managers of research and development, marketing, manufacturing, and support are the "Green Guys." As such, these individuals have direct influence over individuals and groups working within, or below, the third level of the hierarchy. "Green Guys" have immediate, firsthand knowledge of the work being performed and are in a good position to judge its progress and quality. Each week, the "Green Guys" would report the status of task completion for their respective groups. We gathered for this purpose every Monday afternoon at 4 o'clock for the duration of the program, including the first few critical weeks of the execution phase. If a "green guy" was not able to attend, he would send an alternate, so there was always full representation at these meetings. The near real-time data gathered was fed into the computer system to identify potential problem areas. Graphs similar to the one described in Figure 6.1 were produced. Where they indicated a potential difficulty, the appropriate "Green Guy" and I would synthesize a course of action to head off any milestone slip as we made resource and model adjustments. From my perspective, it was a systematic way of focusing the "right" amount of attention on the "right" potential problem area at a time when team anxiety was high.

To complement the tools and the quality and quantity of information they were capable of producing, I found it desirable to create a number of simple reports that would further utilize the work of each manager and team member. These reports were described in the previous chapter. The following discussion explains how they are used in the execution phase.

Individual "To Do" Lists.   Each team member receives an individualized report that lists all of his or her activities ordered by scheduled completion date. For each task on the "to do" list, scheduled start and finish dates are given. Fields for actual start and finish dates are provided. Using the list becomes a simple matter of recording when work on each task has started and when it has completed. The individual does not have to provide any other data. Accordingly, completing the "to do" list may take a few minutes a week. By proper arrangement of the columns on the sheet, a graphical indication of progress to plan builds with the passage of time. People learn quickly that they will receive far less managerial attention if their work is tracking close to plan. Individuals ahead or behind plan can expect a visit from their immediate supervisor and possibly from other managers in the hierarchical structure to discuss the situation and develop modes of action for correction.

For the individual, the "to do" list provides a view of what has been accomplished and what remains to be done. It becomes the communication vehicle between the person doing the work and the management team. When changes to plan have to be made, this list serves as the means by which the change is input to the computer system. As the system processes the modified data, an updated "to do" list is generated. Manager and contributor discuss changes and alternatives and work to reach agreement on the revision. Agreed-upon revisions are accepted only after determination of their impact on the entire program. Therefore, no local decisions are made in a vacuum.

Managers usually keep a copy of each of their employee's "to do" lists. With this information arranged in the proper form, it is easy to set team goals for weeks or months in the future. A manager can request that a "to do" list be printed for his or her department. Taking this action shows which tasks as a function of time the manager is held accountable to have completed. Ultimately, the summation of "to do" lists leads to a task listing for the entire program sorted by scheduled finish date—a report called the scheduled completion report.

Scheduled Completion Report. This report sorts all activities by scheduled completion date, providing visibility of upcoming activities. It is used primarily by the program manager and the other managers at the second level of the hierarchical tree—the "Green Guys," as we dubbed them. This group uses the report to determine if task completion is on target and where problems exist. Furthermore, this management team uses the report to gain an understanding of what tasks need to be completed within the next week, month, and quarter. It becomes their view into the project's future.

Review of the scheduled completion report is done at each weekly "Green Guy" meeting. In effect, the information content of this report sets the agenda for the meeting. A typical agenda follows:

Compare this week's actual progress to plan
    Identify problem areas
    Discuss solution alternatives
    Take responsibility for action
    Look into the future
Identify what needs to be done in the next week
    Discuss potential problem areas
    Begin thinking about solution alternatives
Identify what needs to be done in the next two weeks
    Discuss issues that may limit progress
Identify what needs to be done in the next month
    Discuss issues that may limit progress

Use of the scheduled completion report in this manner gives the management team a complete cross-functional view of progress, problems, and future actions. This is the precise recipe needed to make rapid, integrated decisions that keep the execution phase moving to its timely completion.

Schedule Exception Report. There are two parts to this report. The first lists activities that were scheduled to start but have not, while the second lists activities that were scheduled to be completed but are not. The team members responsible for each of these activities are listed too. The report alerts the management team to potential problem areas. Additionally, sort criterion utilized to generate this report can be used to produce the graphical summaries displayed in Figure 6.1.

Alphabetical Task Name Report. This report is generated to help you remember what something is called. For those individuals that interact with the tools and the database, it is an indispensable item. As the database grows in size, the value of this report grows in direct exponential proportion.

Conducting weekly management meetings that make good use of the information generated by the new product development team is an effective means by which to transition into the execution phase. Managers must determine how each of their people is responding to the change and how effectively their teams are performing to plan. Without doubt, serious problems will surface. For product development teams employing this methodology for the first time, dealing with the reality of the situation will be a challenge. However, it is highly unlikely that the entire plan will have to be redone in the first month. Usually, only a few areas will be far removed from predicted performance. First, the problems will have to be isolated and then solutions will have to be developed. Fortunately, troubleshooting and problem isolation are relatively easy.

First, determine if the network is correct. Second, decide if work quantity and work commitment figures are in error. Third, determine if the right resource is assigned to the task. If the network is correct, proceed to step 2. If it is not, make the adjustments in the impacted regions of the database and then proceed to steps 2 and 3. Managers and contributors will find themselves discussing models, estimation data, actual data, and resource assignments to arrive at a solution to accelerate project progress. This is precisely the dialog that needs to be taking place in the areas that are experiencing problems! Changes to model and database need to be produced in near "real-time," most of the time within a few hours and usually never more than a day or two, to keep plan and project in step—welcome to the execution phase.

6.2   GETTING THE JOB DONE

As the new product development team begins to gain some understanding of reality, a steady-state condition starts to emerge. Managers go to their weekly meetings; participants work on their tasks. As they report on progress and problems, data get collected, analyzed, and reported, and problems get identified and resolved. The process begins to work in a uniform, steady, and, most of the time, predictable fashion. To provide an indicator of the new product development system in motion, the KMET chart developed in Chapter 5 becomes active once again.

As mentioned previously, the KMET chart was designed to serve three important functions. First, in the planning phase, it provides an excellent measure of the plan's structure and detail in a single-page summary. Second, during the execution phase, it tracks key milestones over time to measure slip against the original plan. Third, plan structure is recorded over time to give an indication of the growth and change in complexity of the project/program. After becoming familiar with the chart, a user will be able to extract this information at a glance.

Milestone slip is easily read from the KMET chart shown in Figure 6.2. The baseline plot shows the predicted milestone completion dates with the solid arrows. The running plot shows actual milestone completion dates with a dashed arrow. In the case of project XYZ, the "Design Review" milestone was reached four weeks

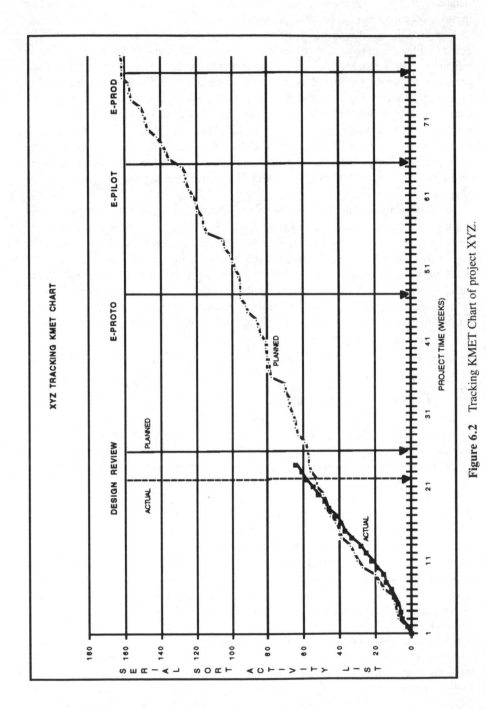

Figure 6.2 Tracking KMET Chart of project XYZ.

204

ahead of the planned date. Milestone slip is the primary parameter under the control of the project/program manager during the execution phase. The manager and his or her team should do everything possible to keep the achievement of milestones on schedule. Frequently, adjustments in the network are required to achieve this objective. Many textbooks on planning refer to the execution phase as the "tracking" phase. If tracking were the only responsibility of the project/program manager, then this would be a time of rest and relaxation! In reality, "tracking" is the time for real-time synthesis of solutions to problems as they are revealed over time. The actions taken to solve the problems show up as modifications to the network from which the KMET chart is produced. The dynamics of network modification and adjustment to the realities of actual task completion dates is shown by the "Actual" line in Figure 6.2.

Comparing "actual" cumulative task completion to "planned" cumulative task completion in Figure 6.2 shows that the project got off to a slower start than anticipated. As time went on, the entire team was able to adjust the rate of task completions per week to a slightly larger value. This slight change in slope held steady so that by week 22 the team had caught up with the schedule of the original plan. At this point, they returned to a winning position with respect to plan. The completion rate held constant for a number of weeks after crossing the planned line; this enabled the team to complete all work necessary to conduct the design review four weeks early. With the first milestone being reached four weeks early and the project's task completions per week establishing a slope greater than planned, project XYZ looks to be headed in the direction of early completion if no major problems surface. It would be nice if all KMET charts would look like Figure 6.2; unfortunately, this is not always the case. What would a KMET chart look like for a project that is having difficulty making forward progress with respect to plan?

The appearance of a change in plans over time is a phenomenon that is highly likely. The KMET chart provides the window into the project/program network to keep change measurable and visible. At any point during the execution phase, it is simple to tell from the chart how much the plan has changed since the reference plan was produced. The chart prevents projects from terminally being in the planning phase and not the execution phase by giving the project/program manager the freedom to modify plans to keep milestone slips under control. *This is real-time, seat-of-the-pants project/program management in action!*

Once the new product development team learns how to adjust plans in real time, the KMET chart takes on new significance. It shows the organization the amount of change the project/program had to absorb and the resultant slip in schedule. For those of you who deal with the changes without allowing a major milestone slip, the KMET chart demonstrates to the team (and those observing the team) the versatility of the people performing the work. When slip does occur, the KMET chart can also be used to show justification for it. Additionally, it will present the new plan of attack and the resulting impact on milestones. The links between task quantity, task completion, network, and time become observable to the entire team.

With each passing day, tasks get completed and the measurement of progress is used to focus energy on future work and problem resolution. The execution phase control system remains in full operation until the last major milestone is achieved.

6.3 PROBLEMS SOLVED/MINIMIZED

Tracking the product plan and taking appropriate action during the execution phase can help the new product development team focus on and solve the problems that have been found to have a negative impact on project/program performance. Refer back to section 1.3 for the list of process problems and discussion of their causes to refresh yourself about those items commented on in this section.

Organizational Instability. It is normal for an organization to change over a period of time. Often, upper managers (officers, general, functional, etc.) will move to assume new responsibilities, thereby creating an opportunity for another aspiring individual. When this happens, the organization's strategy is frequently affected. Managers who are new to the job can benefit from reviewing the definition and plans of a project/program that is in the execution phase. If these elements are supported by concrete evidence of progress (the KMET chart and reports help here), the new manager can make a quality decision on the level of support he or she will provide to the project/program. In effect, a stable definition supported by a documented and tracked plan of execution can serve to stabilize organizational strategy in the event of management movement. On the other hand, a new observer finding a project poorly defined, marginally planned, and missing key deliverable dates is likely to attempt to remedy the situation by taking the project back to the starting blocks. Project managers have some influence on their destiny when upper management changes occur by keeping the project in control and by having evidence to support their assertions.

Upper management turnover effects may be small in comparison to problems that are produced by changes in management responsibilities on the existing project/program team. A complete task network that is sorted by team manager is of great value in this situation! It contains an item-by-item list of work that the manager's work group is responsible to perform. Knowing job responsibilities makes it a relatively easy matter to identify, recruit, and install a replacement. The replacement manager (who the departing manager can help be brought up to speed with minimal assistance) has access to information on all the task activity going on in his or her group sorted by the individuals responsible for the work. In effect, an up-to-date plan that is accurately tracked makes it easy to move managers into and out of the project/program.

Another subtle effect of organization change will sometimes be that the product's definition, or original plan, is slightly modified to accommodate new management desires. When this happens, the effects of the change can be added to the plan and results can be communicated within the organization. The ability to forecast the impact accurately makes it all the more valuable to have a solid base from which to draw comparisons. Take note that the project reference plan along with key reference milestones is plotted on the KMET chart. Most of the time, small changes that do not affect major milestone completion dates will be accepted, whereas those that lead to major delay will be rejected.

People Turnover. Usually, a project manager does not know who will leave or when this will take place, but it's reasonably certain it will happen at some point. If normal courtesy is followed, they will have up to two weeks to deal with the problem (other changes can happen more suddenly). In general, this is a problem that cannot be backed up with a contingency plan because a surplus of human resources is not available. It is, however, a problem that can be quickly resolved (frequently without milestone impact) with the aid of the individual activity assignment report.

Keep in mind that this report lists, for each individual on the project/program, all of his or her remaining activities ordered by scheduled completion date. It can be thought of as a job description. This report provides the manager with information to correct the problem in a number of ways that are appropriate to the particular situation. One alternative would be to shift remaining work onto someone who is ahead of schedule and has available time. A second would be to bring in a replacement from another department in the organization. Third, the manager might find it desirable to hire new talent. With the individual activity assignment report, it's clear to see a reasonable way to work with the problem; yet many organizations decide to deal with this situation without this information.

Inexperienced Staff. Chances are the project's manager will be working with at least some, if not many, inexperienced individuals in the execution phase. The team may include individuals straight out of college, those just hired from another company, or those who are simply new to the job at hand. Their professional backgrounds and training may vary considerably. Managers must handle this diverse group of individual talents properly to gain effective job performance. As discussed below, the results of the definition and planning phases can be used during the execution phase to deal with these problems.

The definition can be used to communicate the reason the project is being done and the general direction the team will follow to achieve its objectives. To those new to their job, knowing where they fit in and why their work will make a difference really helps. This information can contribute to generating that "can do" spirit within the team. By using a common motivational vision of the future, the project manager can help build enthusiasm, dedication, and teamwork. In so doing, he/she will set the stage for having their team focus on the tasks at hand, building experience and expertise that is beneficial to the execution phase, while discouraging actions that may not lead to direct productivity enhancements.

The plan and, in particular, the activity assignment reports, serve to establish the expected output from each individual as a function of time over the execution phase. This ordered list tends to communicate to the team member the exact nature of the job. With the activity assignment report, what must be performed and when it should be completed becomes apparent. For someone new to their job, this can be the road map to success as they track their progress against the plan and share the results with their manager.

Individually, or with management cooperation, the individual can look for ways to accelerate work that is lagging behind plan. The impact that late completion of one task has on others can be directly measured. Corrective actions can be taken to keep work on schedule. This cumulatively produces a win–win situation by placing focus on the "right" task at the "right" time.

In certain cases, it becomes clear to the individual and the management team that the appropriate assignment of individual/resource to task has not been made. The existence of a good plan facilitates the necessary adjustments. Early detection of task delay (in most cases, it can be observed in a matter of a few weeks) enables these adjustments to be made before the scheduled completion date of major project/program milestones is impacted.

Technology Changes. Improvements in time-to-market (TTM) performance tend to assist in the elimination of this element as a delay factor. The reason is obvious—get the project/program done before a change can take place! For any industry, it is possible to gain an insight into the rate of technology turnover as a function of time. Once these times are reasonably well understood and forecasted into the future, they can be used to set TTM goals. When TTM moves faster than technology change and faster than competition, the project stands to complete the execution phase without delays produced by a technological advancement. However, improved TTM is not a complete guaranty that this will happen.

Once a project enters the execution phase, it usually cannot absorb a major technology change, but some minor changes can be incorporated that often make a significant contribution to the marketability of the product. The distinction between major and minor is sometimes not clear. To understand the true implications, it is necessary to change the definition and the plan.

The plan that is in effect in the execution phase can be used to test the impact that will be produced. The new technology can be studied and its method of incorporation into the product can be synthesized. In effect, the team will do a small repeat of the definition and planning phases. The information provided will lead to the synthesis of an updated definition and an updated plan. A comparison can be made between the plan that is presently being executed and the revised plan to determine if the schedule delay can be offset by increased product sales performance in the marketplace. If it makes good business sense, the new plan will probably be adopted and the new technology will be incorporated into the product.

This discipline tends to eliminate the "wouldn't it be nice if" people from having a negative impact on the execution of the program/project simply because new technology just became available. These individuals tend to focus on technology for the sake of technology and often lose objective market focus. Whenever possible, this form of thinking should be kept under control during the execution phase. The definition and planning phases help to control this situation.

Design Errors. Mistakes happen! Some are worse than others. Some occur randomly; others can be predicted if properly observed. For obvious reasons, errors in system design tend to have a greater impact on project milestone dates than those in hardware and software modules. The methodology described in this chapter can assist the project team in dealing with random and predictive errors in design.

When an error is detected, the appropriate person, or team, goes into the synthesis mode to figure out a way to deal with the problem. Usually, the detection of the error is reported well ahead of the solution sequence. Project managers can practice a form of management by wandering around by getting involved with firsthand discussions on the error and potential solution methods. Then, with the appropriate design people, they can create and insert a task network into the plan. Milestone impact can be measured to perform another iteration on the error-correction network. Once agreement is reached, the updated plan can be communicated and tracked. Tracking and understanding the cause of each error can assist in differentiating random errors from predictive errors.

Once a source of design errors is identified, it generally can be corrected. Design tools can be modified, replaced, or abandoned. Design processes can be documented and analyzed to isolate zones of weakness. Error-prone people can be trained and coached. In a systematic way, design errors can be brought under control. This zeal to get at the root cause speeds execution of the present project/program and yields lessons that can only benefit future projects that will enter the design process. It is only through systematic project tracking that potential areas of concentration for improvement become visible.

Testing Reveals Problems. Having testing expose a number of problems with the product during the execution phase is almost a certainty. Allowances for this effect should have been made within the plan by adding test activities into the structure of the network and by establishing task durations that anticipate test problems. In cases where this foresight was not part of the plan, project managers can expect to do a fair amount of real-time management when the problems are revealed.

Once noted by a team member, the problem can be recorded and an assignment made to an individual, or team, to develop a solution. Where the solution is quick and easy, little needs to be done except to acknowledge that the work has been performed and to make certain that no major milestone has been impacted. When solutions are not readily obtained, managers get involved in helping to establish a plan of attack. This plan can then be added to the product execution network and tracked through to completion. Therefore, it is reasonable to expect the project/program network to undergo modification each time a major problem is uncovered. In exercising this discipline, problems are prevented from "falling through the cracks" and a log is automatically maintained to produce data that can be used in future projects to estimate the amount of time to allocate for the resolution of problems on a per-unit basis.

In projects/programs in which problem resolution is factored into the base plan, project managers should follow up on each test problem as it is reported to make certain it is within the range structured into the plan. As input is received that goes beyond the anticipated range, the task of solving the problem can be assigned and added to the structure of the project/program network. With this action, the problem is automatically recorded and its solution is now an integral part of the tracking system. Problems that have been revealed by testing can be assigned a special code to assist the team in isolating particular problem areas.

Unforeseen Tasks. Task omission is most frequently caused in the planning phase by individuals who lack a clear vision of what to expect in the future and/or are overoptimistic. As was stated previously, many teams are composed of inexperienced personnel, and wishful thinking, may obscure reality, so it should be expected that most any plan will contain major omissions. Managers have to deal with unforeseen tasks in each project/program undertaking. If this is so, then they should expect to see the number of tasks that their team must perform grow as a function of time. This, in turn, keeps the project/program manager actively involved in updating the project's network plan as new tasks are discovered and added.

Experience and data from major applications of the methods contained within this text suggest that the preceding scenario is true. The KMET chart in Figure 7.5 had the number of tasks grow by an approximate factor of 2 over the duration of the program. For this program, the growth in task count occurred because I did not have the vision necessary to predict the structural detail of the PERT 12 to 16 months into the future. Contributing to my lack of predictive vision was the limited vision of the team members who contributed to the plan. They also saw less to do in the future than they saw closer to the present.

Once individuals have gone through the predictive process in detail, they'll be better prepared to anticipate such phenomena. Experience is the best teacher. Only by recording task structure on present projects is it possible to create a history for future projects and build the experiential elements into the new product development team culture. Memories from previous projects and their historical databases become a means by which a team can view a new project's future.

Each project/program that follows the methodology described in this text leaves behind a task network that contains detailed information that can be used by future project teams. Utilization of such a database can significantly reduce the task growth on new projects in the execution phase. The importance of keeping the database is that this information can now be transferred, with little training or effort, to new teams. Managers and their teams can use the data to become better synthesizers of work structures and to make better time estimates.

Dependent Project Delay. The network model of the execution phase makes the scheduled delivery of output from one project to another a highly visible event. If the transfer is on time or within the available float of the dependent project, the timing of the overall execution phase will not be affected. On the other hand, if delivery is late, the project's schedule may be negatively impacted.

With the model in place, it is no longer necessary to guess about what will happen and what should be done. First, determination of when delivery can be expected becomes the highest priority of the day. Obtaining the necessary information may require active participation, discussion, and investigation to isolate cause-and-effect relationships. Commonly, technical professionals and managers will have to interact to create a number of possible alternative delivery scenarios. With each scenario defined, it then becomes possible to determine to overall impact on the dependent project. Networks are modified, resources are reallocated, and decisions are produced. The final determinants of corrective action become the model and the information that drives it.

Customer Requests Change. The new product development effort is usually well into the execution phase before customers become very aware of the final product's existence. Therefore, a request for change at this point in time can be a rather emotional event for project members. It's something that is likely to excite the project team, but just what should be done? Sometimes making last-minute revisions pays off. On the other hand, they are often unnecessary and foolish endeavors when done without knowing the basic business implications.

Once the request is fully understood, the network model is used to determine the true cost of making the change. Most likely, the request will require the structure of the network to undergo substantial revision. As network modifications are produced, resource redeployment scenarios are simulated to staff the newly created tasks within the network. The individuals making the adjustments will become fully aware of resource needs and the delay in market entry that will be produced by making the customer-requested revision. Generally, the cost in development dollars is small compared to the revenue dollars that will be lost because of a late market entry. Thus, the delay determined by network analysis is used to project the effect on revenue for late market entry. Essentially, the managers of the new product effort can use the information at their immediate disposal to rapidly arrive at the return on investment of the requested change, taking into account the primary factors of expanded development costs, delayed market entry, and the projected gain in revenue that results from making the revision. At no time should a change that increases cost and delays market entry be entertained without some understanding of the impact on sales.

Based on the numbers, one of three possible decisions will be reached: don't make the change, make the change at a later date in the next revision of the product, or make the change. Obviously, the first two alternatives have no immediate impact on the execution phase of the new product development effort. The third alternative, while requiring extended effort, is no longer a major obstacle for the team; in the process of preparing for the decision, they have already determined how to make the change. Once the "go" decision is given, the team just has to implement the change. This is done by making the revision to the model a full cross-functional exercise and then allocating resources to the newly created tasks within the network.

User Requests Modification. During the early stages of the execution phase, this phenomenon rarely occurs, but as a prototype of the final product is made available to customers, people who come into contact with the product begin to form strong opinions about how it should work, how it should look, and how it should be controlled. As more replicas of the product are produced and more people are exposed to the form, fit, and function of the design, more user-requested modifications will be generated. With the passage of time, the number of products multiplies so that more people from across the organization begin to interact with the product, and, in turn, the number of users exposed to the product increases. For some, the interaction will be casual; for others, a significant percentage of their time will go into thoroughly understanding the product. Both the casual and in-depth user can be expected to suggest ideas for improvement. The more exposure each has to the product, the more critical their evaluations and comments are likely to become. Neither can be ignored and a response to their inputs must be generated.

A fine balance must be struck between allowing product developers the freedom to make improvements without unnecessary managerial intrusion and making management authorization of any change a necessity. This delicate balance is achieved by using the project's model as an indicator of effective attention to user-suggested modifications. Data and analysis will determine what level of control is necessary to keep the project progressing at its scheduled pace.

When response to user-requested modification is out of control, most everyone associated with the product development effort is attempting to make an enhanced contribution that goes beyond what is really needed. The program manager tracking progress is likely to see a substantial number of individuals falling behind schedule, thus jeopardizing the planned completion dates of tasks. If the model is structured as recommended, the situation is likely to be detected within two to four weeks. Usually, detection within this short period of time enables control of the situation to be regained. The manager will need to discuss options with the individuals involved and develop an alternative approach to reestablish forward momentum. In these circumstances, the highest priority of the manager is to establish firm guidelines regarding acceptable and unacceptable team response to user-requested change.

On the other hand, if no task completion dates are missed and some product enhancement activities are taking place, it is probably best to allow individual contributors the freedom to make reasonable improvements without managerial intervention. This is a sign that informal cross-functional paths of communication are working and that team participants know that enhancements are welcome as long as the overall project retains its forward momentum.

At times, certain user-requested modifications may be rather substantial and, in the judgment of the management team, necessary to implement. In this case, the model must be revised so that work is effectively distributed among participants and so that the entire team understands the full implications of the decision and the product changes involved.

Shift in Market Requirements. As has sometimes happened, the market shifts while the product is in the execution phase. Upon detection of the event, three actions are recommended: (1) assess the situation; (2) based on the best available information, quickly decide what must be done; and (3) make the necessary adjustments and proceed.

The concepts presented in Figures 1.1 and 1.4 can help the management team assess its present situation. How far is the project in the execution phase? If it's still early, cumulative project expenses may still be relatively small in comparison to what they are to become. Also, it may be possible to make a major shift in product implementation strategy and not have a major impact on delivery of the product to market. Time on the team's side affords it the opportunity to better determine cross-functional cause-and-effect relationships than if it was in a rushed mode of operation. On the other hand, a team that has crossed over the 75 -percent point in expenses and will have a product on the market in a short period of time is likely to view the situation from a completely different perspective.

While developing a thorough understanding of the situation, the team will also be working on alternative courses of action that could be used to counteract the shift in requirements. It's possible that some, or maybe even all, of the team's activities will be focused on doing definition phase type work for a brief period of time. Typically, the time needed to prepare the alternatives for consideration is relatively short—a few days, maybe a week. Adjustments to the product's definition will be produced as the cost in resources and time are determined for each alternative. The team and its managers interact with the existing project model to determine what will change, impact on resources, effect on development cost, and shift in project time to market. Fusing information relevant to impact on development, and projecting benefit derived by adjusting to the shift, lead to the production of key information that will be used to decide on a recommended course of action.

Decision makers now have within their possession information on development timing, resource utilization, and added development costs along with the data on projected market benefits derived from making the shift. Data presented to the decision makers will help explain the benefits of not making a change vs. the results that are likely to be realized by pursuing one of the several alternative courses of action. In effect, decision makers will be working with both internal and external information to arrive at a decision. Decision makers and development team participants work together to make a well-informed decision in a very brief period of time. The decision thus determines future action.

Obviously, if it's decided that responding to the shift is not worth the additional development costs or projected loss of sales caused by delayed market entry, then the project continues with its original plan intact. Naturally, adjustments will have to be made to compensate for the effort that went into definition- and planning-related activities that were needed to fully understand the impact of the shift in market requirements.

A decision to respond to the change results in a full and thorough update of the model. Individuals and managers of the cross-functional product development process need to contribute their input as the plan undergoes a significant revision. The act of using a cross-functional, shared replanning process serves to explain to the team what is being done and why it is being done. Communication channels within the organization are fully utilized as participants plan for the change. As the new model and plan materialize, it is run through the same careful analysis procedures outlined in Chapter 5 to ensure its quality.

Once the replanning effort is complete, the new product development effort, with its goals modified, returns to the execution phase.

Unanticipated Competition. When unanticipated competition emerges, a competitive business analysis is in order. In an alternative analysis, the cost of continuing the execution phase (based on an accurate definition, plan, and measured progress to date) can be weighed against the threat posed by the competitor's product offering. Having followed the methods within this text gives the management team a much better footing on which to base future action.

More than likely, a product effort that has gone through the rigorous definition and planning effort outlined in the text will need to make few, if any, alterations to the execution phase plan to remain in a strong competitive position.

Changes in Marketing Plan. Examples of change to a marketing plan include such items as a change in the feature set of the product, a suggested revision in the performance specifications, a dissatisfaction expressed about the color or appearance of certain product attributes, expression of disagreement over differences between product simulation as compared to present new product reality, and straightforward "gut feel" criticism of the product with no supporting evidence. Essentially, changes in the marketing plan are suggested subjective changes to the product's documented and undocumented definition that appear late in the execution phase and that an individual, or small group, thinks will greatly improve the marketability of the product.

The number of suggestions made and/or acted upon increases the more product samples are available and the closer the actual product is to customer delivery. Generally speaking, these conditions are present in the last half of the execution phase. It is usually at this point in the process that marketing professionals get their first full view of the product that they are expected to promote and sell. Before this point, the image of the product was set by simulations, artwork, models, and test fields. Most of the time, what these individuals thought the product would be and what it is are very close. There will probably be areas, however, where the reality of what has been produced differs substantially from the mental image held by the marketers. Even with the existence of the external reference specification (ERS) and good-quality simulations of the final product, some areas of suggested change are likely to surface. Individuals in marketing will be strong proponents of change because they perceive that making the revisions will make their jobs of introduction, promotion, and sales easier. Other members of the team will resist the change because it is just more work at a time when there is already too much to do!

Once again, good judgment and responsible balancing of different perceived actions need to be exercised. The project's database can be used to determine specific cause-and-effect relationships produced by each suggested change in direction. A critical screening parameter becomes the trial assignment of the revision to a resource and the determination of the impact of task load addition on the resource. If the data manipulation by the tool and the response provided by the assigned resource suggest that no schedule delay is likely to be produced, then it is likely that the change will be scheduled into the regular flow of work. In situations where there is a significant impact on the new product's time to market, rigorous attention must be given to the request for change. The main question that must be answered is, Is making this change at this late stage in the process really worth it? The more that is known about the actual costs in terms of time and money, the higher the probability that the change will not be implemented. Only those revisions that can be shown to offer relevant payback will make it through the tight filter established by the methodology. This is precisely what must be done to maintain focus and objectivity as the execution phase nears completion, and uncertainty about the product's performance in the marketplace builds.

Leadership Methodologies. Leading the execution phase of the product development effort on a cross-functional basis requires individuals with special talents. People who feel they have to micro-manage projects are not suitable for the challenge. Individuals who think it is necessary to

Schedule work for their subordinates down to within a few hours per activity per day;
Know who is on vacation and who is sick as they keep track of the information on a computer system;
Collect daily status reports from each individual contributing to the effort; and
Be actively involved in each and every decision made on the project

are probably not well suited to lead a new product development effort. At the other extreme, people who

Work only by personal interaction;
Keep little or no data on the project;
Don't know where trouble is likely to develop and why; and
Seem to be chasing after the project to stay with it

are no better qualified to lead the effort than those previously described.

An effective leader will fully understand the ideas behind the material presented in this text and be able to make the methodologies work. This person will set up the system and utilize it to develop the new product effectively. Use of the information provided by the system keeps the individual targeted on the problem areas of the project and prevents the waste of his or her precious time on areas not needing attention. By observing the data on the project and interacting with the participants, the effective leader determines where to concentrate resources while making the necessary implementation decisions to keep the effort focused on meeting its end objectives.

As the execution phase proceeds, the leaders of the effort must be able to recognize certain signs of distress. Some questions to keep asking, categorized by phase, are

Definition

Is the ERS stable?
Do development participants understand the product?
Does the market seem to be expanding?

Planning

Do large portions of the original plan remain?
Has the organization delivered on its resource commitments?

Execution

Are personnel asking to stay with the project once the planning phase is done?
Is milestone slip under control?
Is the customer satisfied?
Is management well removed from day-to-day leadership of the effort?
Are technical problems under control?

Whenever an answer other than yes appears, adjustments must be made to the effort. Only a leader utilizing the methods described in this text will be prepared to take timely decisive action.

Acts of God. Such unexpected events don't often happen, but when they do, it's nice to have a base upon which to determine future action and upon which to assess impact.

6.4 BEYOND THE EXECUTION PHASE

When the execution phase is over, four activities should follow in a timely and quality fashion. First, the new product development team should take time to celebrate their achievement. Second, members of the team that can help ensure a successful manufacturing and sales buildup should be provided the encouragement and opportunity to do so. Third, time must be taken to review the entire new product development process, as well as the data that was generated, so as to learn from the experience. Lastly, knowledge and experience from the present success must be applied to future projects. Some detail on these four activities follows.

A new product development team that brings a product to market has produced a significant achievement. Of all the ideas that people have on product development alternatives, only a select few survive and are delivered to the market. Having a new product in existence serves as a catalyst to initiate a period of relaxed behavior and celebration.

Thought should be given to the best way to conduct the celebration. An awards dinner with the entire team and their families and friends in attendance is often appropriate. At the dinner, managers and development participants can present awards that have some serious and some humorous content. The celebration might be conducted in a "roast" format, in which many individuals share in the give and take about particularly noteworthy times on the project. During the course of the evening's fun, the top management of the organization should offer its congratulations and express its appreciation to the team.

In advance of or during the team party, a project memento should be distributed to the participants. Though usually an inexpensive item, it is likely to be of immediate and long-term significance, and therefore serves an important function. Mementos can be simple, such as coffee cups, project photo albums, T-shirts, or sample pieces of the product's technology molded into plastic. These are the kind of things an individual can share with family and friends to describe the development environment and the contributions made by the product. These artifacts stimulate individuals to think about what they did, how they did it, and to remember some of the pleasant aspects of the new product development adventure.

Top management of the organization and leading managers of the product development effort need to recognize the team and key individuals in public and private fashion. On the public side, articles in the company newspaper should talk about the new product and the people that produced it; press releases, data sheets, and advertising materials should be readily available to make it easy for team members to share the work they have done with others; and community and civic links should be established between the product and the people that helped in its creation. On the private side, key contributors may be selected to receive wage increases, bonuses, stock rewards, or a combination of the three. Where monetary compensation is provided, it should be closely linked to the behavior that led to the effort's success, not to the position the recipient occupied or the normal responsibilities that position entailed. In this way, such rewards become relevant and an incentive to all team members.

Lastly, team participants should have every available opportunity to interact with customers as the product goes to market. This can be accomplished in subtle ways. Engineers can be encouraged to attend trade shows where the product is on display and to lend a hand with "booth duty"; research and development personnel can accompany marketing and sales professionals when conducting customer demonstration and application qualification sessions; and support staff can be involved in in-house tours that are given to customers visiting the organization. With a little creative thought, it's usually possible to bring just about everyone associated with the new product's realization into contact with end users. Doing so builds a sense of pride and satisfaction within the organization and demonstrates to others that new products are considered to be very important.

As the period of celebration fades, the team's attention shifts to helping launch the product from a marketing and manufacturing perspective. This is an aspect of the new product introduction process that is only briefly addressed in this text. Suffice it to say that plenty of work remains to ensure that the new product will be successful in the market. Some task items to consider are training of the sales team, developing application literature, conducting training classes for customers, moving from small-scale to high-volume manufacturing, and making small improvements to the product as preliminary customer feedback is received. The important thing at this point is not the assembly of a list of all the things that need to be done, but the recognition that people who brought the new product this far may well be the best-qualified individuals to help with its marketing and manufacturing climb to mature unit volumes. Members of the team that bridge the gap between factory and field are certain to gain a perspective on overall new product development process issues—one that will surely return large future rewards as the entire process repeats.

As work continues on making the new product a market success, the entire process that led to the creation of the product needs to be studied. Assuming the team followed the methods outlined in this text, data on the definition, planning, and execution phases are in ample supply. Since the output of each phase is specified, analysis simply means examining it in view of the total process of creating the new product.

For the definition phase, attention should focus on corporate and division definition documents, present and future new product definition sheets, and current strategic recommendations. Based on what was learned, some changes should be made to the documents and the process that leads to their creation.

A frequent discovery made after following the process outlined in this text is the fact that the business has too many new products in its new product development portfolio for the amount of resources it has available. The second most frequent discovery is understanding how to measure the time to market of a product. With data in place, it becomes clear that the measurement starts with the conceptualization of the idea and its first representation on a new product definition sheet (NPDS) until the time the product actually appears on the market. This somewhat obvious fact of measurement is not fully appreciated until organizations document the length of each phase and begin to see that the definition phase can sometimes extend far longer than the execution phase. In other words, moving on ideas quickly after they appear may be the key to competitive or early delivery of product to market. Determination of the significance of these items lies within the process data. New discoveries are certain to result.

The product definition process is certain to improve. Data now exist to significantly amend the estimation process throughout the entire new product development organization. Each function now has assembled data on the complexity of the end product that was produced and can determine the productivity of the work groups that contributed to the process. Each function can easily synthesize curves that resemble those shown in Figures 3.3 through 3.6. No longer will it be necessary to guess at relationships that link work complexity and productivity to the time required to complete the work. Data on change to the external reference specification (ERS) will also have been recorded over the duration of the project. Each change can be viewed on an individual basis and the entire list of changes can be collected, categorized, and analyzed. Understanding the mechanisms that led to change in the ERS is certain to lead to improvements in future definition documents.

With the completion of just one planning–execution cycle as outlined in the text, future planning phases will never be the same! The reason is simple—the organization is now in possession of accurate data on minimal serial networks, task work content figures, and actual hours worked on the tasks. In addition, it becomes possible to begin the development of statistical figures associated with particular tasks and resources. No longer will anyone preparing a new plan have to guess what the network should be or what work quantity figures should accompany a particular task. In effect, new plans are likely to be based on significant amounts of well-researched and recorded data from projects that preceded them. Future planners will only have to take planning risks in areas where no previous data exist. Typically, the percentage of future plans based on incomplete network and task data will be less than 5 to 10 percent of the overall plan's content.

Data on the execution phase can be used to make further enhancements to concurrent engineering projects throughout the organization. Identification of problems and the recording of the data that led to their resolution are certain to benefit future projects and teams. The mystery of the execution phase no longer exists. Organizations grow from individual and collective perspectives and thereby become better at what they do. Participant selection and training now become matters of task identification and linkage to areas of expertise within the organization.

When all possible lessons have been gleaned from the planning–execution cycle, it becomes time to begin the application of data and experience to future new product development efforts. Making this possible simply becomes a matter of having people move on to their new assignments and making raw and summarized data available to all future projects. As each new effort moves through all phases, experience and data build, making it easier for each subsequent effort.

7

APPLICATION

In this chapter, you will learn about

The use of portions of the methodology on an actual project.

The definition of the product and its complexity.

The method used to synthesize the plan.

The KMET chart of the project.

The problems encountered during the execution phase and the actions that were taken to resolve them.

The performance of the various teams on the project.

Task dynamics that had major impact on the network.

The development of the "Survivor's Index."

Viewing the methodology as a control system.

The possibility of finding a universal tool to solve all development problems is remote. However, by careful observation and reflection you can assemble a kit of tools capable of solving most problems.

Preston G. Smith and Donald G. Reinertsen

7.1 INTRODUCTION

This chapter's focus is on the development of the Hewlett-Packard 8791 model 10 frequency agile simulator. It's probably fair to say this product development effort helped shape my future and that I, in some small way, had an influence on the product that was delivered to market. Certain aspects of creating this machine were far different than any I had experienced on previous projects.

I began working on the product concept in the early 1980s, when the idea was just a twinkle in the eyes of very few individuals within the Hewlett-Packard organization. In particular, a man that I admired and respected, Rolly Hassun, was behind a concept that was so fresh and challenging that it was impossible for me to go about my assigned duties without thinking about the power contained in the ideas he was presenting and proposing. Rolly and a close working associate, Al Kovalic, were breaking down political and technical barriers that had previously not been attempted. Through their personal effort and the foresight of the management team within the division at the time, people and funds were allocated to grow investigation phase staffing and begin the development of key technology parts at sites in

California and Colorado. I joined the effort as R&D Section Manager at the time the first product architectures were being defined and technology investigations were being initiated.

Just about all definition work for the product concept was led by individuals in research and development, with some key support being provided by product marketing champions. From the start, this product concept was defined for the market by R&D with marketing participation. As you will see, some aspects of having an R&D-driven product definition worked satisfactorily; others, whose effects were not revealed until the execution phase, did not. Those individuals in R&D who advocated pushing the product to market suffered later on as a result. We had to invent definition phase concurrent engineering practices as they were happening. As we discovered after the project was over, we were dealing with scope and complexity issues that were new for just about everybody in the organization.

To the best of my knowledge, the HP 8791 was the first product development effort in the company's history to utilize 3×5 inch cards on a cross-functional basis to develop a shared plan of the execution phase. Having this plan made it possible to guide the development of the product using first-of-a-kind methods that were completely within the character of the total undertaking. If we had not broken new grounds with planning methods and tools, we certainly would not have been able to guide the execution phase through the problems that surfaced as well as we did!

Lastly, the execution phase was a managed by the numbers task that provided key feedback on the system that was created to conduct the execution phase.

The presentation that follows focuses on what was done, how it was done, and the results obtained from the application of new cross-functionally integrated methods and tools. Many of the concepts presented in earlier chapters were inspired by my participation in this adventure. You'll certainly be able to recognize the linkages between theory, experiment, and application. A description of the product follows first, to give you an understanding of the magnitude of the job under consideration. The models that were developed for the project are then described. Discussions on definition, planning, and execution follow. At each step, examples of materials that were utilized are presented for observation and comment. Finally, the data produced during the execution phase is analyzed.

7.2 PRODUCT DESCRIPTION

A predecessor to the 8791 model 10 was the 8770 arbitrary waveform synthesizer. Some basic operating theory on the 8770 helps to explain the concept behind the 8791 model 10.

An arbitrary waveform synthesizer (AWS) is a device that lets you recreate signals that have been generated on a computer but whose speed of point-by-point playback is far faster than the operating speed of the computer. Therefore, the signal is constructed and then stored within the computer's memory, but it has no way to get out and do anything useful. Each point on the waveform is represented by a 12-

bit binary number. This sequence of numbers within the computer's memory is transferred to high-speed buffer memory contained in the 8770. With the numerical representation of each waveform point in the synthesizer, the operator of the product specifies the rate at which each numerical point is supplied to a digital-to-analog converter for transformation to an analog signal.

Conceptually, the compact disk player can be used to describe the operation of the arbitrary waveform synthesizer. The sequence of binary numbers that is used to feed information to the digital-to-analog converter is stored on the disk itself. The rate at which data are presented to the digital-to-analog converter is set by the rotation velocity of the disk and the digital information on the disk. Playback is then accomplished by reading the numbers from the disk, processing them, and finally supplying the processed data to a 16-bit digital-to-analog converter to recreate the audio signal you hear. Usually, about 40,000 digital words representing points on the analog waveform get processed each second the compact disk player is in operation.

Instead of using a disk to store the digital representation of the signal, the 8770 uses multiplexed high-speed memory for waveform storage. This memory is the type used in some of the world's fastest commercial computers. Its speed of operation differentiates these memory devices from those used in lower-performance personal computers. The existence of fast memory implies that a processing and conversion system perform at least as fast. In the case of the 8770, its speed of operation is so fast that it is possible for the system to convert 125 million, 12-bit words into their analog equivalents each and every second.

In 1987, the 8770, a first-of-a-kind product, created a new market for high-speed baseband signal simulation. It offered performance that was not commercially available, at a reasonable, but high-profit, price.

As a stand-alone product, the 8770 was of limited value without supporting tools. It would be like having a compact disk player with no disks to put in it! To make the product as useful as possible to a broad, but highly specialized, audience, the Waveform Generation Language (WGL) software package was produced. This software tool provided users of the 8770 the capability to create the waveforms they wanted on their computer system. In effect, they used their computer workstation to design the waveforms they would later reproduce with the 8770. In comparison with the compact disk, WGL made it possible to create the information that goes on the disk. This capability was certainly necessary to realize initial sales, but it certainly restricts utilization of the product to only the select few individuals that have the desire and talent to program waveforms.

The 8770 and the WGL became the market probe for the 8791 model 10. We were able to promote the concepts of computer-generated waveforms and high-speed digital-to-analog conversion to a broad audience. The 8770 was possibly the best market research tool we could have created for the 8791 model 10.

As we helped guide the 8770 into the hands of pioneering customers in the marketplace, attention was directed to finalizing the definition of the 8791 model 10. With each sales gain or loss on the 8770, we were learning about what we had to do on the 8791 model 10. We were balancing concept, performance, features, development funds, and time to market. At the time, we knew far more about the capabilities of our technology than we knew about how to tame the technology. Development of the 8791 model 10 definition concurrent with the introduction and sales buildup of the 8770 gave us the information needed to get our product's definition into about the right form. As you will read later, we had a good definition phase; it was not perfect.

Basic operation of the 8791 model 10 frequency agile signal simulator (FASS) is described by the equation

$$\text{Output signal} = [1 + A(t)] \sin [2\pi F(t) + P(t)]$$

Those of you familiar with signal communication theory will recognize the equation as that of an amplitude, frequency, and phase modulated sinusoidal carrier. If the equation makes little sense to you, think of it as a way to mathematically describe the signals received by your AM radio, FM radio, or television receiver. The $A(t)$ term makes the signal get larger and smaller as a function of time. The $F(t)$ term modifies the carrier frequency of the signal that is somewhat analogous to an FM radio transmission. Color on your TV receiver is produced by proper adjustment of the $P(t)$ term. Thus, by proper modification of amplitude, frequency, and phase terms over time, it becomes possible to send information from point to point using radio waves. Where this theory was once simply applied, modern device technology and increased spectral utilization have made it possible, and also necessary, to increase system complexity to further serve our needs. In this environment of complexity and efficient spectral utilization, the frequency agile signal simulator makes its contribution.

FASS combines direct digital synthesis and agile up-conversion to generate advanced signals for designing and testing modern systems. It is the off-the-shelf building block for high-performance simulation of radar, electronic warfare, and communication signals. From a performance perspective, the system

1. Covers the 10- to 3000-MHz frequency range.
2. Can change to any frequency in its range in less than 250 nanoseconds.
3. Creates simultaneous $A(t)$, $F(t)$, and $P(t)$ over a 40-MHz modulation bandwidth.

The system is shown in Figure 7.1. Examination of the illustration shows the system to contain four subsystem units and a CRT/keyboard interface. Two of the system boxes contain the equivalent of three 8770 arbitrary waveform synthesizers, each responsible for amplitude, frequency, and phase modulation of the carrier; digital signal processing subsystems; and the digital-to-analog converter. The third unit translates the output from the digital-to-analog converter to exist at any frequency in the 10- to 3000-MHz frequency range. Lastly, the fourth unit is a special-purpose computer that integrates the operation of the system through its human interface (keyboard and CRT) or its machine interface. We referred to this fourth unit as the "smart interface."

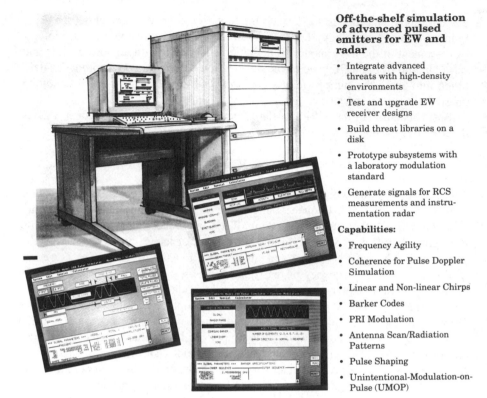

Off-the-shelf simulation of advanced pulsed emitters for EW and radar

• Integrate advanced threats with high-density environments

• Test and upgrade EW receiver designs

• Build threat libraries on a disk

• Prototype subsystems with a laboratory modulation standard

• Generate signals for RCS measurements and instrumentation radar

Capabilities:

• Frequency Agility

• Coherence for Pulse Doppler Simulation

• Linear and Non-linear Chirps

• Barker Codes

• PRI Modulation

• Antenna Scan/Radiation Patterns

• Pulse Shaping

• Unintentional-Modulation-on-Pulse (UMOP)

Figure 7.1   Frequency agile signal simulator (FASS).

It was the "smart interface" and its software concept, called the Instrument-on-a-Disk (ID), that made FASS very easy to understand, program, and operate. We learned from working with the 8770 and WGL that many sales were not being made because the product combination was just too hard to use for many potential customers. Instrument-on-a-disk was created to reverse this situation. It gave the end user a control panel that could be used to adjust software parameters tailored for specific applications. In effect, the Instrument-on-a-Disk (ID) made the unit work more like a compact disk player works—you merely insert the disk you want to hear (ID application itself) and select the track you want played (ID control panel). Once selected, FASS makes the "music"! An example ID interface screen is shown in Figure 7.2.

From a complexity perspective, FASS consists of

• More than 11,000 electrical components.
• Three custom-integrated circuits. .
• Three specially engineered radio-frequency components.
• More than 25 custom components developed by outside suppliers.

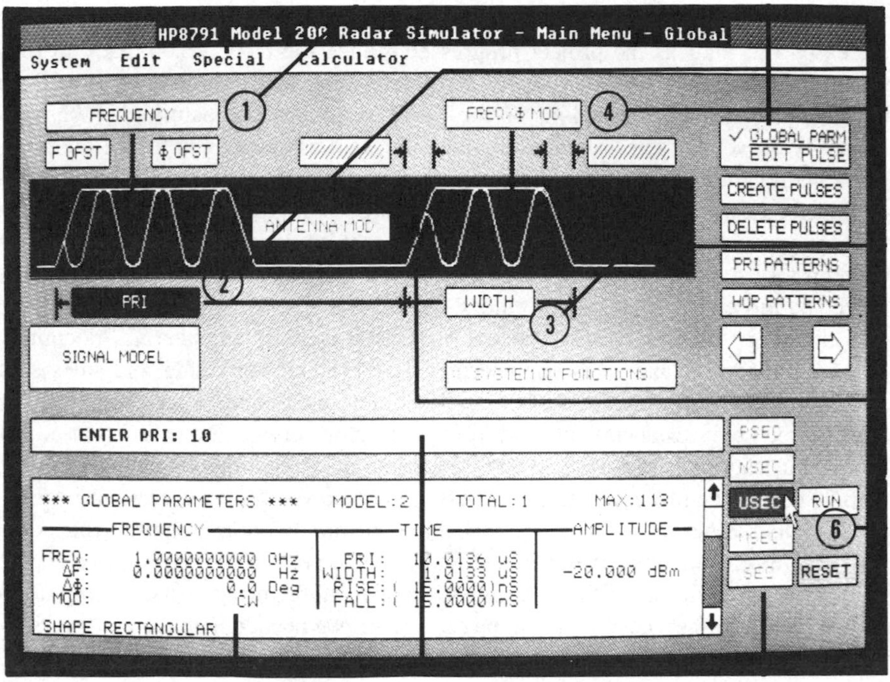

Figure 7.2   Example Instrument-on-a-Disk screen.

- Greater than 40,000 lines of firmware.
- Almost 100,000 lines of newly developed software.

Production of a product with the complexity figures of FASS was not a trivial undertaking. While doing the program, I often thought of the products I had developed that fit into one package and performed a stand-alone function—they were so much easier to do!

7.3   PRODUCT DEVELOPMENT TEAM

In research and development, the undertaking was segmented into hardware, software, and systems functions. One project manager was assigned to develop the two instruments that contained the digital hardware and subsystems, another to lead the development of the up-converter, and a third to create the smart interface. In addition, the project manager who was responsible for "smart interface" hardware was also responsible for the Instrument-on-a-Disk (ID) and system software. Shared responsibility at the system level was also part of our R&D team structure. Approximately 25 engineers were assigned to this team and distributed between the hardware, software, and systems functions.

Manufacturing had assigned team participants early in the definition phase and increased this staff as the project progressed. Since the 8770 was already in full-scale manufacturing, it was easier for us to determine what would have to be done differently to simplify operations on the FASS system. Manufacturing's focus was primarily on vendor selection and qualification, and assembly and test of the product. Qualification efforts were led by manufacturing materials engineering; development of new assembly processes and techniques was the responsibility of the fabrication and test team; and manufacturing test engineering had the charter to develop test stations and test procedures for the products. In all, this group was staffed by more than 15 key contributors.

Marketing was divided into product marketing, service engineering, documentation engineering, and sales engineering. The product marketing and sales staff were about the same size. A service and technical writing team was assembled to create three key user documents and to fully develop service and calibration procedures, illustrations, software, and text. Once into the execution phase, we had more than 20 marketing professionals contributing to the development success of FASS.

Of course, other individuals were needed to ensure that other requirements could be met. Quality and environmental testing had to be performed. Our finance people worked closely with us in the determination of manufacturing costs, pricing alternatives, and recommended financial targets. Also, we needed help from our personnel department. In all, our support staff numbered about four professionals as the project moved from milestone to milestone.

When you add up the cross-functional staffing numbers, the total is in excess of 60 professionals. The complete cross-functional hierarchical structure of the FASS project is shown in Figure 7.3. The complexity of information flow and linkages between cross-functional teams on this project was, by far, the most intricate of any project I had ever directed. Understanding the cross-functional hierarchy and knowing the basic interaction that would be required between the teams had a major impact on my approach to this leadership and management challenge.

With the functional hierarchical structure defined, it became obvious what the next step had to be—our team needed a common milestone sequence to follow. Fortunately, our working environment was rich in milestone sequences. There were a number of milestones defined for projects at division, group, and corporate levels. All we had to do was pick and then emphasize the ones that made the most sense for this particular project to follow. We selected four of the most common milestones used in the division. They were design review, laboratory pilot (LP) review, production pilot (PP) review, and introduction.

Design review is a point in time within the execution phase where measured data on prototype hardware and software are presented to upper management to secure their support for the LP phase. Before reaching this point, prototypes of all hardware are produced; sufficient software and firmware exist to operate each system component individually; and some system software exists to measure fundamental system parameters. If the design is approved, the project enters the LP phase.

Laboratory pilot (LP) designates a phase in which the product's design is replicated a number of times to verify that it is manufacturable. To verify manufactura-

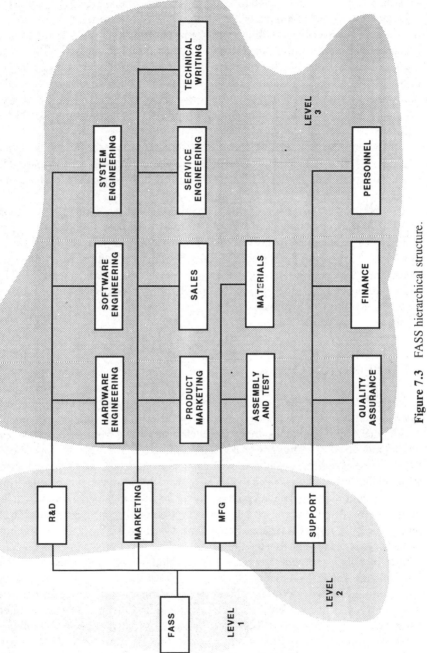

Figure 7.3 FASS hierarchical structure.

227

bility, however, may require the involvement of people outside of the manufacturing function to make the product and processes function. Unfavorable interactions are identified and corrected through modification of process on product design.

With a number of LP products successfully built and thoroughly evaluated, attention is directed in the production pilot (PP) phase to verifying that manufacturing processes can produce the product with desired performance and within targeted costs and times. As such, the PP phase is one that measures process performance. Individuals outside the manufacturing function contribute very little to the effort in this phase. On the other hand, systems software begins its final integration and test once PP units have been produced. Quite often, PP units are shipped and sold to customers.

Lastly, introduction activities relate to formal announcement of price schedules and delivery times. In this portion of the execution phase, products are demonstrated and promoted to generate orders and, hopefully, some backlog for manufacturing.

Combining the information in the preceding paragraphs should make it easy to understand the matrix model of the FASS project shown in Figure 7.4. The model has functions defined on its horizontal axis. Phases are defined on the vertical axis. For FASS, the primary phases of definition, plan, and execution were followed. The execution phase was further segmented into the design, laboratory pilot (LP), production pilot (PP) and introduction phases. Review of the time axis in the execution phase shows the broad overlap and the concurrent nature of the execution phase segments.

7.4   DEFINITION PHASE

A feasibility model of the system's digital electronics and signal processing subsystem was built in a relatively short period of time by a few engineers early in the development of the product's concept. Much was learned from this exercise. In particular, device complexity, power consumption, and preliminary performance limitations were revealed. The more performance designers tried to extract from the system, the larger it grew and the more power it consumed. Also, operating the digital system at high speeds revealed many important parameters to consider in final system design. Likewise, knowledge was acquired on the critical design parameters of the digital-to-analog conversion system. This was a true technology-driven first-of-a-kind effort—the initial investigation team had thoughts on potential applications as it proceeded to make the machine perform the functions it thought would be needed in the market.

This preliminary system, and the information building it produced, was used to promote the product concept to secure organizational support for its continued development. Observing the "PARSYN" (its early code name) perform in public excited many engineers and managers. This initial machine could do that that was impossible to produce by any other method. As knowledge of its capabilities expanded, financial and staffing support for its continued investigation and technology

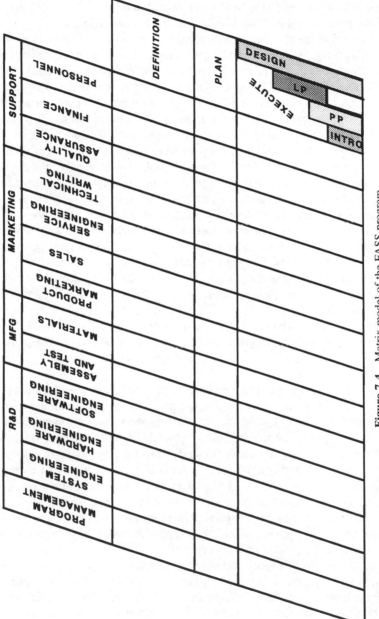

Figure 7.4 Matrix model of the FASS program.

development followed. Work on the digital-to-analog converter took place within facilities in California, Washington, and Colorado. Digital signal processing integrated-circuit work was done between California and Colorado. System design, individual instrument definition, and other related definition phase activities took place within the Stanford Park Division in Palo Alto, California.

Over a period of years, work was directed toward getting the technology to perform. Making a high-speed, spectrally pure digital-to-analog converter was an extremely difficult technical challenge. Initial schedule estimates were found to be in error by factors of 2 and 3. The integration of integrated circuits into their packages and the integration of packages into subsystems became major engineering efforts. Concurrent with this work, design and fabrication of the digital signal processing integrated circuits were taking place. It turned out that these devices were much easier and faster to design and produce than other aspects of the system. Thus, we had them available for design into the final instruments well ahead of the digital-to-analog converter. As integrated-circuit performance figures became better understood, final subsystem and system design parameters were being established.

Concurrent with technology development, a small effort was under way to understand the potential market for the product. To this day, it is not clear how much of the hard work we did with our potential customers was of value. The reason I say this is because I don't think many of them understood what we were talking about! However, it must be said that those who reviewed our early concepts and ideas were very kind and hospitable. This preliminary customer contact was inspirational but provided little concrete evidence on which to base design decisions and to make engineering trade-offs. Therefore, the product effort, while slightly modulated by market input, remained technology- and "champion"-driven. The market had no pull on what was being done. Pushing concept and product to market was the standard mode of operation.

As is usual in many large organizations, we also had our share of political problems over this time frame. Quite frankly, we probably spent more time on internal political issues than we did on market issues throughout most of this early period. As it turned out, a split in operations geographically divided the people on the technology development team. This split resulted in two facilities trying to develop and differentiate product offerings that were to serve a market that each of them knew practically nothing about. Much of what I learned about the differences between established and new markets was a direct result of my involvement in this exercise. Needless to say, knowing now what has been said in this text about first-of-a-kind efforts, we never really accomplished anything useful. Eventually, the decision was made to concentrate the entire effort in one facility.

As final development of key technology yielded marketable performance, we were faced with a funding shortage. Rather than slow the entire FASS effort, we made the decision to staff the 8770 team fully and introduce the model before introduction of FASS. This action got the 8770 to market in record time to give our technology and concepts a true customer test. As expected, the feedback provided by customers on the 8770 became very valuable in fully defining FASS and ending the definition phase.

Adapting the FASS definition to what we had learned from the 8770 became a way of life for a period of time. With each 8770 sale, we learned about the strengths and weaknesses of the product concept. The learning experience was reflected in the product definition of FASS undergoing substantial revision. Because of the generic nature of the hardware, most modifications to the FASS definition had major impacts on the software development effort. The bulk of feedback received regarding the 8770/WGL combination indicated that it was just too difficult to use. This customer feedback led to the creation of the Instrument-on-a-Disk concept as the entire software focus changed from one where we were just going to upgrade the Waveform Generation Language (WGL) to one where we would provide complete applications with solutions.

Effectively communicating the concept of IDs and the "smart interface" to potential customers now became easy, since it was an alternative they understood. We were now in the position to demonstrate differentials in features, performance, and price between a number of alternatives. As a result, the quality of information received from end users improved by orders of magnitude. The software engineering team developed simulations of IDs that gave the proposed screen image of the product. Customers and users gave us their feedback. We became quite good at adjusting the ID and "smart interface" definition to accurately reflect the feedback. We considered various hardware alternatives that each ID could support and the performance that could be realized. There was a period during which open lunchtime classes, referred to as "Digital Synthesis for Lunch," were held to communicate concept and receive feedback on alternatives. Through a process of comparative analysis, agreement on a preliminary FASS definition was reached. The "vision" of what needed to be done began to look very similar across functions and within the vertical structure of the organization. If we had had a definition matrix, we would have been able to accurately measure our progress. Unfortunately, this tool had not yet been invented.

The concept of the definition matrix was created as work on the development of this text began. Since it is now in our common vocabulary, it will be used to help explain the state of readiness of the FASS definition. To begin with, FASS had the equivalent of the new product definition sheet (NPDS). In fact, we had four sheets! The two digital instruments, the agile up-converter, the "smart interface," and the overall system were documented with one-page definition summary sheets. Review of these documents shows they were quite accurate in the detail presented. Weakness can be seen in the vision, big picture, and market information cells. Strengths were clearly present in the understanding of technical details and their key attributes affecting hardware, software, and system operation. The major omission made by the FASS definition team was in the production of hard output. An external reference specification, as defined in this text, was not produced for the system. However, external documents for the hardware units and for some of the key software modules were generated. The information that held the system concepts together was stored in the minds of our systems and software people. Reliance on mental imagery at this stage of development worked, but it certainly is not a recommended operating procedure. As you will read later, not having spent more time doing a thorough ERS of the system could have cost us a great deal of expense and time. It was the talent and "can do" of our team that compensated for the weak performance in this key area.

However, operating from our position of definition matrix ignorance, we started to feel good about where we were headed. It was now time to thoroughly explore what steps had to be taken to get there.

7.5 PLANNING PHASE

It was at this point that I thought of using 3×5 inch cards to structure the program. While handling a pack of cards during a card game, I noticed they could be arranged on a table in the form of a project network. I realized that cards requesting the needed information could be distributed to the entire project team, allowing them to join in the planning effort. An example of the cards utilized was shown in Figure 4.3.

The cards encourage people think in network terms (What is the task called? How long will it take? What comes before it? What comes after it? What is the next major project milestone? What else should be communicated?) without having to study the details of network theory. People find it easy and natural to do. In fact, my team of planners found it so exciting that it created almost 500 cards when it first used the idea!

Buoyed by this tremendous response, I set out to arrange the cards into a huge network on the floor. It was at this point that I realized why trying to view projects through, say, a 13-inch diagonal CRT does not work! Actually, it became quite enjoyable to work with the network and experiment with how to best arrange the program for maximum effectiveness. It became obvious that, for this program, a functional structure was the correct answer.

In performing the preceding activity, another problem became evident. The quality of the input from some individuals was well below the minimum acceptable level. It was clear that one-on-one and team training would be needed. Thus, a short course on project management basics was created, though there was not much time to prepare it or teach it. I developed a training system that made it possible to improve people's skills in less than one half day of interactive training. Emphasis was on task structure by deliverable, estimation techniques, series and parallel network diagrams, and hierarchical structures. Response to the training was favorable and rapid. Skills improved to acceptable levels for most players in a matter of hours.

With the quality of input now well above the minimum desired level, a final network was created by each function. The people responsible for production of the final 3×5 cards and I discussed each of their cards as we looked at the network on the top of a table. We used this forum to determine the adjustments that would be made prior to inputting the data into a computer system. To do this required that I interact with the program's cross-functional management team. In retrospect, the time spent on these one-on-one interactions established the relationships that continue to this day. We were learning how to work as a finely tuned team—this was the first step.

Once the discussions were complete, I checked each card to make certain it contained all the information needed by the project management software we had selected. After this last check, the task of inputting the data into a computer system began. In a matter of hours, the system was doing its numerical analysis. "What if" games were played to arrive at the first plan.

The network complexity was reduced down to 216 tasks. Three major milestones (design review, lab pilot [LP] review, and production pilot [PP] review) were predicted and communicated to the entire team. Additionally, a private meeting was held with the general manager and functional managers of the division to explain the process and the results that were produced. Our team had its first success. We had the focus we needed and our motivation to meet, or beat, the plan was in place! We knew a major change had occurred that would make the future exciting.

Next came the task of producing our first KMET chart. The baseline on the KMET chart, as discussed in Chapter 5, is the cumulative task completion vs. time of the original plan. The results of our initial planning efforts are shown in Figure 7.5. It predicted design review would be held in week 9, lab pilot review in week 36, and production pilot review in week 56. Only hard work, some luck, and the passage of time would prove how well we had forecast these milestones.

As you examine the "original" collection of data points in Figure 7.5, recall the discussion associated with Figure 5.5. Can you identify the problems resident in the original FASS plan? A comparison with "actual" and "original" should help you

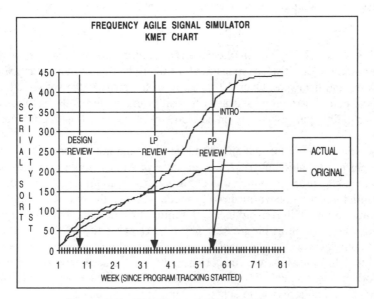

Figure 7.5   KMET chart of the FASS program.

identify the key problem areas. Of course, our team was still at week 0 and we were in possession of the best plan ever produced for the program. Our confidence and "can do" spirit were high. Besides, we just plain did not understand how to interpret what the KMET chart revealed. Thus, we left the planning phase behind and headed into the execution phase commiting ourselves and our organization to the milestones set by the plan.

7.6 EXECUTION PHASE

No project/program attempt within this particular HP division has had this much history recorded about its planning and execution details! Effort was made to accurately track all tasks by recording actual start and actual completion dates. This simple entry provided information on task start, slip, and task duration expansion/compression. Additionally, it became clear that network changes (the addition, removal, and rearrangement of tasks) would be a normal part of the execution phase. To keep a record of the network modifications, plans were saved and dated before modifications for the next period were produced.

During the execution phase, it was apparent that completing this program would require real-time synthesis of corrective action steps. Note the plot of actual task completions of the final KMET chart in Figure 7.5. It shows that the number of tasks grew to 443 from the original 216, resulting in a significantly different task vs. time profile over the duration of the program. The growth in tasks was primarily because additional tasks became evident as one approached the actual work. There were also an abundance of secondary reasons for this, which are identified and discussed below.

Changing Product Definition/Internal Miscommunication. Our software team had a large job to do. It was responsible for the production of more than 130K lines of software and firmware. Over the duration of the project, the software team was the critical path. Team progress was very visible and could be managed to keep the program running quite efficiently on all fronts. The team dealt with a fair amount of change!

Difficulties with product definition and internal communication resulted from not doing a perfect definition phase. Many of these problems would have been avoided if we had had the product column of a definition matrix to follow. On the other hand, a substantial effort had been made to produce what we thought was a high-quality definition. No one will ever produce a definition of a product that does not change as the development effort progresses. The important quality of the definition is its relative completeness. If it's stable enough to facilitate planning the execution phase, the definition already meets a minimum level of acceptability. For first-of-a-kind product development efforts, moving the definition much beyond the minimum may not be all that necessary to achieve overall introduction success.

There is a documentation balance that must be achieved between too little and too much. Compounding the situation is that many people consider themselves to be a design "experts" once they have seen something to criticize. The only way to get to these people early is to give them something visual, functional, and interactive to review and record their feedback before starting the final design. Of course, under realistic product development conditions, that's like saying you have to be finished before you can start. So what could we have done better?

First, we should have listened to our product's cross-functional "champions" more and to our nit-picking "experts" less. We set up a filter, but it was not very effective. Over 80 percent of all requests were implemented—a percentage that is just too high. A simple test on requests for change that made an attempt to measure potential program delay as a function of the knowledge base of the individual responsible for the input would have been a much better screen for us to implement. Second, we were in desperate need of having a few key customers to work with who could have helped us focus on only the really important facets of the product's definition. This could have been easily accomplished if we had directed our energy toward establishing key relationships rather than trying to build analytic market models. Our organization's fixation on doing "me-too" and next -generation products had hurt our ability to better define a product like FASS. Third, we could have done a much better job with the external reference specification. Had we known better, we could have taken action on all three areas. I'll comment in more detail on the ERS.

Where concepts are new, it is probably better to err on the side of doing too much. Improvements could have been made by adding more detail to the ERS in two areas:

1. Human interface.
2. Machine interface.

This indicates that input/output (I/O) must be thoroughly defined and controlled. User screens should have been defined down to the last detail. Control names, control functions, control shapes, control colors, and control locations required detail specifications to minimize miscommunication between teams on the program. This is not to say we needed to create a voluminous stack of documents. I estimate we missed being right by about 20 detail screen descriptions (graphic models) and 80 written pages of text on control functions. The work content of this undertaking would have been about four person-months. Based on the talent of resources available and the productivity of this team, we could have produced the Revision 0 copy for cross-functional distribution and feedback in under two weeks! I'm certain that this time would have been recovered later in the FASS program, so the net impact of improving the quality of the ERS would have been positive with respect to our ability to get the product to market. Also, cross-functional communication channels would have had a corresponding improvement.

The work on better screen and control definition would have made the task of documenting the machine interface rather easy. There is an almost one-to-one mapping between button and GP–IB (General Purpose–Interface Buss) machine commands. Moreover, system control and data communication parameters could have been specified in greater detail. This planning effort would have consumed about two person-months of work and its physical execution could have been accomplished in a week.

My best estimate is that this program was less than a calendar month shy of a "just right" ERS. Such an ERS would have helped us in software development, design, and test, as well as in technical writing, service engineering, product marketing, and sales. Clearly, we were not the first product development team in history to experience this difficulty. What prevented us from having the required documentation in place?

System and software engineering did not receive sufficient input from product marketing, sales, service engineering, and technical writing to effectively define their information needs during the definition phase. The generation of the ERS was not a shared experience. Therefore, as needs became known, the information was exchanged on an informal basis. A period of time was then needed to determine what to do about each piece of data as it surfaced and became understood. Usually, tasks were added and the network structure was modified to adjust for the change. In this way, we were able to get by without the "just right" ERS and still effectively run the program. As you will see, we were able to function with product definition problems and keep delay to a minimum.

People Turnover. Three members of the original management team moved on to new roles within HP during the execution phase. Only one of the three remained associated with the program. There were turnovers of engineers in R&D, manufacturing, and marketing. Each case was dealt with independently to determine future needs and take the appropriate action. As mentioned previously, the task list sorted by individual (the computer-generated "to do" list) was invaluable in adjusting to these changes rapidly and efficiently.

Inexperienced Staff. Addition of new people and modification of assignments to adapt to change were quite routine on the program. Again, the individual reports helped us track progress, provide training when needs became clear, and make adjustments to match each individual's performance level. The team concept played an active role here as well. People would assist others more readily if they knew they were slightly ahead of schedule in their own work.

Design Errors. Designers proved once again that they were human, but this time we were able to incorporate corrective actions into the task network without impact on any milestone. At times, having a slightly larger margin would have been nice! Furthermore, while errors did appear, there was so much visibility on corrective action tasks that it was impossible to have the problem "slip through the cracks."

Testing Exposes Problems. The FASS program had to fix problems that were uncovered during mechanical shock testing and electromagnetic compatibility (EMC) testing. These problems, like almost all problems, were really not anticipated. Previous testing on prototypes revealed that the product was in full compliance with regulations. In the EMC case, the original test was performed in an indoor environment, whereas the final test was performed at an outdoor certified test range. The indoor emulation was not correct. Mechanical problems were induced by slightly modified test requirements. Once the problems were identified, corrective actions were formulated and added to the task network. Both these major problems were corrected without milestone impact.

Unforeseen Tasks. Since this program was the first attempt to implement this new methodology, we had more unforeseen tasks crop up than I liked to count. As they were identified, the network model was modified and resources were assigned to them. In effect, we went through a plan synthesis and analysis cycle each time new tasks were discovered. A few weeks into the FASS program, we could turn major revisions to the network in a matter of hours. Simple changes having few interactive connections could be done in minutes.

Once a historical network becomes available (even if it consists of data from just one project), it's expected that this number will decrease on subsequent projects/ programs that have access to the data.

Again, no unforescen task produced major milestone delay. It was possible to detect the problem in time to apply resources to keep the program moving.

User Requests Modification. A number of revisions, particularly in the software and firmware areas, were produced by internal users of the product who felt the design needed adjustments. A tracking system was in place that allowed individuals to make their requests. A team reviewed each request, considered how it might be accomplished, created a brief (sometimes mental) plan of attack, and measured the impact that implementation would have on the program. Using this systematic procedure, more than seven out of every eight requests were implemented, with predicted negative impact on major milestones. As mentioned earlier, too many of the requests were implemented.

Change in Marketing Plan. Our marketing group had a tendency to be distracted from its goals. They were not able to define their task network structure and maintain its stability. Over the duration of the execution phase, only a small portion of their original task network survived to completion. This situation was accommodated, but produced some friction between various work teams. Again, having recorded the history of what occurred in the marketing group, it would be easier to plan and structure future programs. Additionally, areas for focused training and organizational improvement were identified.

Ineffective Plans. Unfortunately, one team, the customer documentation team, probably worked harder than all the others. It was the team that had none of its original plan survive the execution phase. In fact, its plan underwent four major revisions during the execution phase. This team's role was to produce the necessary customer documentation. Their ineffective planning can be attributed to changes in the product's definition—*a situation that was out of their control.*

We knew we had a problem in this area and knew why. To deal with it, the customer documentation team was moved into the same work area as the software, hardware, and systems engineering teams. In this way, we were able to maintain our informal information transfer network. Engineers and writers were able to interact in the synthesis process. We built a system that could accommodate change and quickly respond to it. It was not ideal, but it worked. As the program approached completion, the customer documentation team was capable of delivering a finished product within two weeks of receiving the final engineering input.

Acts of God. Testing the product on our outdoor EMC (electromagnetic compatibility) test range was delayed, with no impact on major milestones, by rain in September and extremely cold temperatures in February, two highly unusual events for the San Francisco Bay Area which were not factored into the plans.

The next sequence of graphs are provided to help illustrate some of the means by which the data collected during the execution phase can be utilized. You'll also be able to see where the FASS program succeeded with respect to plan and where it did not. Graphs pertaining to the top level, manufacturing, software, and others are supplied. The analysis was prepared by using the network data that make up the baseline on the KMET chart and comparing it to what actually happened over time. The forecasted and actual data were preserved to produce an analysis of this type. Following the analysis is a discussion on testing and an additional section on changes that were made to the plan by various functional groups.

7.6.1   Team Performance: Top Level

Top-level slip (see Fig. 7.6) was held to a maximum of 40 working days over a period of 16 calendar months. Design review,* the first milestone to be reached, was completed three days ahead of the prediction made in the baseline plan. LP review* finished eight days early, while PP review[+] was only three days late. Public introduction slipped by two calendar months over a total duration of 16 months. When calculated on a per-month basis, introduction delay was held to less than 0.5 weeks per month! This introduction delay was the result of late delivery of final tested software and support documentation. The work of hardware engineering, systems engineering, manufacturing, finance, sales, product marketing, and hardware quality assurance was done by the PP review date. At this level, the initial planning methodology (3×5 cards, etc.) was able to predict the public introduction of the product with an accuracy of 12 percent! For a program of this size and complexity, Figure 7.6 serves to illustrate some of the benefits that can be derived by following simple and disciplined definition, planning, and execution methods.

---

*A point in time when the product definition has set. All hardware had been prototyped, key software modules were operational, and elementary system functions had been simulated or demonstrated.
*Manufacturing had built all hardware with assistance from hardware, software, and systems engineering. Most hardware quality testing was complete.
[+]Manufacturing had duplicated all customer deliverables. The product was ready for final system qualification testing.

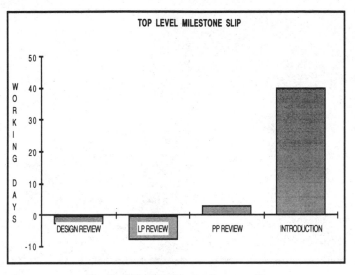

Figure 7.6   Top-level milestone slip.

7.6.2   Team Performance: Manufacturing

Manufacturing displayed top performance in its ability to produce results on time and to predict the duration of tasks. Data from the tracking system make it possible to illustrate the team's performance graphically.

During the LP phase, final assembly of the first three systems started on time (see Fig. 7.7). The fourth started a day early and the fifth, six days early. It's important to keep in mind that the start date prediction for the fifth unit was made about 40 weeks earlier!

Predictions of LP task durations (see Fig. 7.8) were equally impressive. LP1 took 10 days longer than planned, but this extension was recovered in subsequent unit network activities.

As the program moved into the PP phase, manufacturing continued its outstanding performance. Maximum start delay was held to a peak value of five days on PP3 (see Fig. 7.9). Three of the units started within a day of the predictions made well over a year earlier.

PP task durations (see Fig. 7.10) were less than those in the LP phase, but problems with system PP3 and system PP5 prevented the manufacturing team from achieving the textbook performance it had forecasted.

7.6.3   Team Performance: Software

The software team was equally impressive in its ability to forecast task durations, especially when considering that the estimates were based strictly on preliminary firmware and system definitions. We used the estimation techniques described in Chapter 3 and some of the targeting procedures discussed in Chapters 4 and 5.

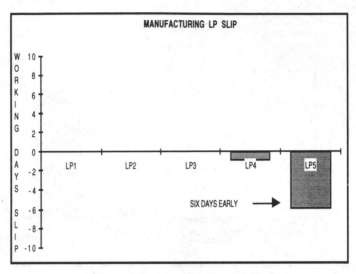

Figure 7.7   Manufacturing LP slip.

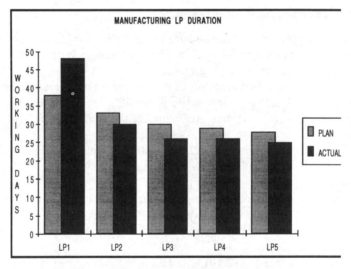

Figure 7.8   Manufacturing LP duration.

Using these methods, the software team was able to predict task durations to a maximum error of 20 percent, with average error held to under 11.7 percent over the duration of the effort. However, software schedule slip followed a more classic pattern.

Slip (see Fig. 7.11) resulted from small increases in task duration and decisions to do unplanned work before starting a scheduled task. As mentioned previously,

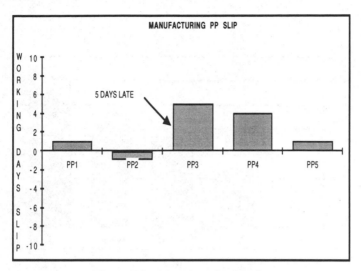

Figure 7.9 Manufacturing PP slip.

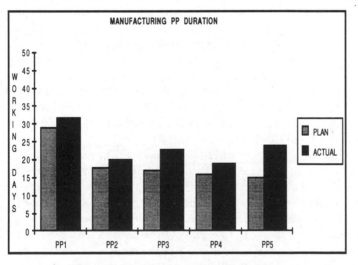

Figure 7.10 Manufacturing PP duration.

the people doing the initial planning in product marketing, sales, customer documentation, and customer support did not effectively communicate their information and service needs to the software engineering team (nor did the software engineering team solicit it). Hence, software engineering had to alter the product definition as needs became known. It was called upon to add and modify features, provide training and demonstration materials for use by product marketing and sales, and furnish detailed information to the customer documentation and customer support teams.

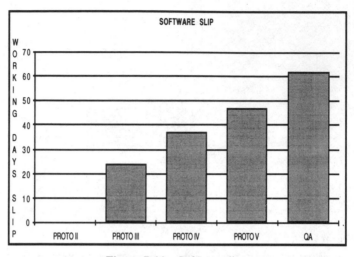

Figure 7.11   Software slip.

Knowledge about the product's functionality emanated from the software engineering group. Release of new information carried with it the possibility of generating additional work.

In terms of software duration, Proto II was completed exactly as predicted (see Fig. 7.12). The selection of the final system hardware delayed the completion of Proto III by only a few days. These decisions had an impact on the software block diagram, module definitions, and software development strategy. These impacts necessitated change in the software portion of the network in order to maintain forward progress. It was an impressive feat to limit cumulative task duration slip to only four days during the Proto III phase.

Most of the design was completed in Proto IV. Effort was directed at completing the interface and math accelerator hardware design, prototyping unit level GP–IB drivers, coding kernel software, and developing the first prototype interface system software running on the system computer. The Proto IV collection of tasks were the most accurate forecasters for future tasks duration expansion. Their rate of completion, as measured on the KMET chart, could be used to predict future performance from this team of software developers. Since extension was held to 10 days out of 60 (16.7 percent) on this difficult portion of the job, the remainder of the plan was structured to run at this rate of slip. For many of us, this news came as welcome relief. We had developed a software development structure that was effectively dealing with large system problems and we had its schedule under control. It is also interesting to note that slip did not start to occur until much of the design was completed and people outside the software engineering group began using the system software to do their work.

Final coding and application development was reserved for Proto V, along with production of the system computational hardware and software for systems PP1

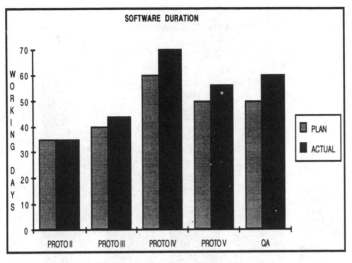

Figure 7.12   Software duration.

through PP5. As can be seen in Figure 7.12, this work took six days longer than originally planned.

Quality assurance testing consumed the remaining time allocated to the software engineering part of the program. On a percentage basis, the team made the largest errors in attempting to predict QA durations. During this time, system performance was tuned and defects that were revealed during test were corrected.

7.6.4   TEAM PERFORMANCE: HARDWARE ENGINEERING, SYSTEMS ENGINEERING, MANUFACTURING ATE, FINANCE, AND MARKETING

Other groups were able to keep their work on schedule and thus from obstructing the critical path, and they demonstrated a reasonable ability to predict task durations (see Fig. 7.13). The hardware R&D team missed the goal by 54 percent. While this seems considerable at first glance, the tasks that expanded in time were not items that interfered with the critical path. Most of the effort went into making the hardware more readily manufacturable and solving radiated emission and shock problems.

Systems engineering predicted its duration with an error of 19 percent. Various factors (determination of system specifications and performance, test generation, hardware integration, software integration, cabinet configuration, manufacturing costs, transportation systems, specification verification, and quality assurance testing) were structured into the original plan. Along the way, systems engineers dealt with a number of problems that could have produced milestone slip. As each situa-

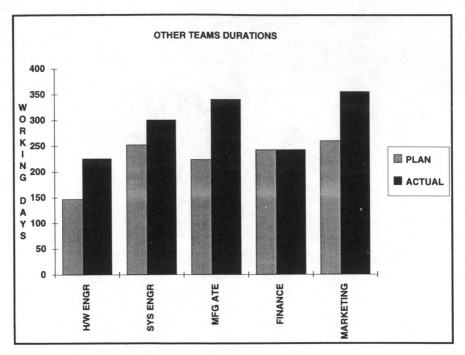

Figure 7.13 Durations of other teams.

tion surfaced, adjustments were successfully made to the network to keep the team off the program's critical path.

ATE (automatic test equipment) is used within manufacturing to verify unit and system electrical performance. Original estimates were based on the upgrade of existing manufacturing test systems to meet the new requirements. For the most part, the strategy was effective. When it became clear that certain performance parameters could not be accurately measured by the ATE, new equipment and new software development efforts were added to the network. This resulted in a 52-percent extension in task duration and necessitated some retesting on units produced earlier in the program.

The network was structured so that the finance team would be primarily driven by inputs from marketing and manufacturing. The inputs were supplied on the scheduled delivery dates, and finance met its forecast exactly.

The marketing data are the collective performance of product marketing, sales, customer documentation, and customer support. Sales team performance, while not scheduled in the original plan, met all scheduled deliverable dates with only slight slip delays and duration extensions. Product marketing had a tendency to stay near the critical path, but had enough flexibility to keep from obstructing it. Without the software delays and extensions, product marketing probably would have had key tasks within the network become obstacles on the critical path.

7.6.5 Team Performance: Quality Assurance

Figure 7.14 provides a view of estimation performance with large errors in forecasting actual durations in mechanical and electromagnetic (EMI) testing. Underlying these time extensions is a network that underwent frequent revision to keep testing from interfering with the critical path. Many creative ideas were conceived to structure what was initially a serial test flow into one with a number of concurrent parallel paths. Total duration was forecast to be 70 days; the actual duration turned out to be 75 days. This was an area that benefited a great deal from timely synthesis of execution strategies, tactics, and logistics to keep the program on track.

7.7   TASK DYNAMICS

Figures 7.11 through 7.14 show how the task count of the network changed as a function of time organized by function. Retention of this data was a natural part of the methodology. Its use as an aid in program/project management was not discovered until the program was past the PP review milestone.

For manufacturing, the profile remained relatively flat except for major revisions in month 4 and month 10 (see Fig. 7.15). Month-4 changes were due to adding module test activities that were overlooked when the original plan was produced. Month-10 growth was caused by expanding detail on the five PP systems from three tasks per system to nine tasks per system based on knowledge acquired during the LP phase. With the addition of tasks to the network, manufacturing management was constantly making the needed assignments of resources to each task. Thus, as

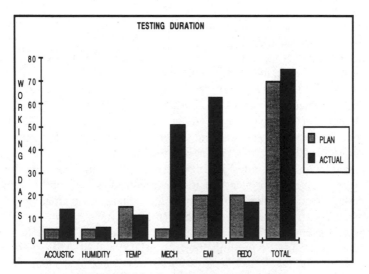

Figure 7.14   Qualification test durations.

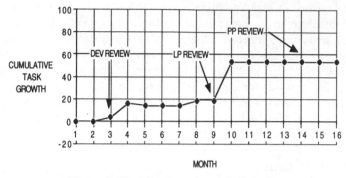

Figure 7.15 Manufacturing task dynamics.

network complexity grew, so did the required commitment of resources to each of the new tasks. As noted in previous discussion, manufacturing had no difficulty associated with the added task count or increased attention to detail.

Hardware and systems activities grew from month 0 through 4 and remained constant to month 10 (see Fig. 7.16). Early growth in network detail was produced by attention to printed circuit board process detail, addition of tasks to address mechanical shock problems, more systems details, and addition of information on LP and PP unit allocations to various work groups. Major changes in month 10 were caused by the program entering the PP phase. At this time, all activities were placed under critical review to focus this team on the last big thrust. Detail was added. Work was redistributed. Adjustments to the original plan were made that utilized the knowledge acquired and that dealt with the scheduled departure of two engineers and a project manager, as well as the addition of one new person to the systems team.

The product marketing and sales profile in Figure 7.17 shows little task count change until month 8. In month 8, the big increase was produced by the addition of the sales network to the overall program network. Before this time, this team's activities had not been included—a potentially serious omission! Additionally, the network underwent four major revisions in months 1, 2, 4, and 6. The growth in month 2 was produced by plan revision. Other revisions did not have a major impact on task count. Task count growth changed by small numbers as new activities were revealed (sales training, beta test revision, family plan, and people motion) and in month 10 yet another major plan revision materialized as the result of moving into the PP phase. Clearly, there was a fair amount of management attention paid in this direction to ensure the production of scheduled output items.

Customer documentation task count grew in month 4 (see Fig. 7.18). This team made extensive plan revisions in months 4, 9, 11, 12, and 14. The decline in month 11 shows the magnitude of the effect of this one revision on task count. In general, this group was struggling with providing a definition of work. Just about every change made to the human and machine interfaces resulted in a task modification

R&D/SYSTEM TASKS

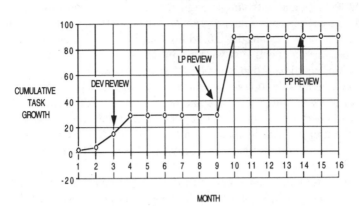

Figure 7.16   R&D hardware and systems task dynamics.

MARKETING/SALES TASKS

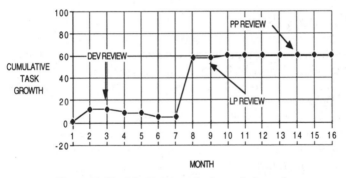

Figure 7.17   Marketing and sales task dynamics.

CUSTOMER DOCUMENTATION TASKS

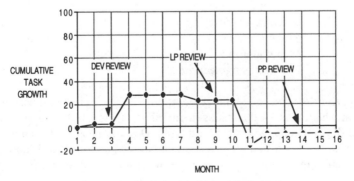

Figure 7.18   Technical writing task dynamics.

within the technical writing group. Moreover, since many items were communicated only verbally and with simple engineering sketches, unnecessary errors were produced. Only at the end of the program was the definition in such form that it was possible to plan and execute the job without change to plan. As mentioned previously, we became so effective at working in this manner that significant documentation revisions could be produced in less than two weeks.

Of course, being able to revise documents this fast is not necessarily the correct solution for all programs. I highly recommend, and I think the people who worked in this environment would agree, making the external reference specification a prototype of the user documents that will follow. In this way, the process encourages discipline in the definition phase and provides a reasonably stable structure upon which to base execution phase plans.

Also, be advised that seeing data of this type captured my attention and the attention of most managers associated with the FASS program. We were doing all we could to stabilize the situation. Only with stability achieved late in the execution phase was our fine team of documentation specialists highly productive on their assignments.

7.8 SURVIVOR'S INDEX

The analysis by task count growth was quite interesting. It showed that *addition of detail* to the original plan had little effect on potential milestone slip, whereas *change in plan* served to foretell major problems. This observation led to the creation of the "Survivor's Index."

This index looks at names of tasks that produced the baseline of the KMET chart. A comparison is made between these original task names and those task names that were actually completed. This is a measure of the stability of the original plan. The results for each team follow:

Planning Team	Survivor's Index
Customer documentation	0%
Product marketing	31%
Software engineering	75%
Systems engineering	77%
Manufacturing	94%
Hardware engineering	97%

The Survivor's Index called direct attention (via low scores) to those areas of the program in which most of the management effort was focused to strive to minimize milestone slip and keep the program on track. Developing an understanding of why the scores are low is a first step in the synthesis of a workable solution. For the frequency agile signal simulator, low scores in customer documentation and product marketing would have directed us to work on the "right" areas to gain an understanding of the definition problems. As the methodology described in this text becomes more frequently utilized, it is expected that this index will assist many managers of new product development processes in identifying potential problem areas.

7.9   SYSTEM PERSPECTIVE

Any system can be viewed as a process that results in the transformation of input into output. Stated another way, the output is related to the input through a system transformation. Systems can be structured by series or cascade interconnections, parallel interconnections, feedback interconnections, and all related combinations. In its most simplified state, the methodology described in this text can be modeled as three closed-loop systems, as shown in Figure 7.19.

The diagram helps focus attention on some of the basic properties of systems. These properties are observability, invertability, and stability. Observability simply means that information (output items) that flows within the system can be seen and measured with some degree of accuracy. Once information is observable, it can be processed in a feedback system to construct a closed-loop system. A system is said to be invertable if by observing its output we can determine its input. This means that the transfer function of each of the blocks is deterministic. In a stable system, a small change in input leads to a response that does not diverge as a function of time. Clearly, a system must be stable to ensure proper functionality over time. In feedback systems, this is sometimes difficult to achieve. It is instructive to examine the procedures and results from the system concept that have been successful.

Observability is evidenced by the flow of output items through the entire process and the system's ability to monitor their production over the life cycle. In the "definition work" block of the system, its output was observed in terms of a complete product description. This output was further segmented into descriptions of output items that were to be provided by each team that participated in the program. The "planning work" block output was observable in the form of the 3 × 5 card network, the baseline on the KMET chart, and the first issue of individual activity

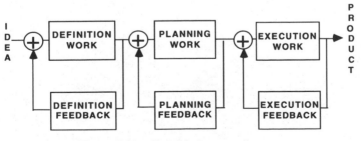

Figure 7.19   Simplified system diagram.

assignment reports from the computerized planning tool. The tracking performed with the "Green Guy" system, measurement of task slip, task duration, and task addition, and the computation of the Survivor's Index demonstrate the observability of the "execution work" block. Once observability of the idea to product life cycle was demonstrated, it was then possible to use the information to control the system.

Output from the "definition work" block was supplied to the "definition feedback" block, transformed, and then added at the input to keep this portion of the system operating at full capacity. While in operation, this output was fed forward to the "planning work" block so that definition information could be used to initiate the planning process. Full interaction with the team of planners through the use of information supplied on the 3 × 5 cards and through "what if" scenarios on the computerized planning system made it possible to synthesize the output from the "planning feedback" block. Tracking the plan and making adjustments through the "execution feedback" block closed the loop on the execution process.

The reader familiar with control system concepts should recognize that the design presented in this book has system theory at its foundation. It is only through observability within a system that its output can be controlled and predicted.

The presentation and analysis of data on FASS are intended to demonstrate the effectiveness of many portions of the methodology described in the text and to show what resulted from their application. Hopefully, they show that the methodology aligns with basic human tendencies. Project/program teams want to succeed while working within a rather simple, easily applied framework! It is anticipated that future utilization and continued refinement will serve to further remove the mystery from transforming ideas into products. In addition, I'm optimistic that the evidence is sufficient to convince you that the idea-to-product process can be defined, measured, and controlled on a real-time, dynamic basis.

APPENDIX

Here, you will learn about

The detailed new product development process (NPDP) matrix.

An example corporate business definition.

An example division business definition.

An example product definition sheet.

An example of corporate objectives.

A model that can be used to help make financial decisions based on the product's time to market.

An example evaluation of the project management software tool "Viewpoint."

NEW PRODUCT DEVELOPMENT PROCESS (NPDP) MATRIX

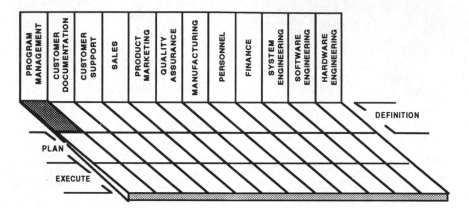

PHASE: DEFINITION   **TEAM**: PROGRAM MANAGEMENT

OUTPUT:

❑ INTEGRATED PROJECT/PROGRAM DEFINITION DOCUMENT

❑ PRODUCT BUSINESS PLAN

❑ PRODUCT PROTOTYPE

❑ PROTOTYPE PERFORMANCE BENCHMARKS AND DATA

❑ ESTIMATED PROJECT/PROGRAM TTM

❑ PRESENTATION MATERIALS TO SECURE ORGANIZATION
"GO AHEAD" AND RESOURCE SUPPORT TO ENTER THE
PLANNING PHASE

SUGGESTIONS/REQUIREMENTS:

❑ VISION STATEMENT

❑ PROJECT/PROGRAM PURPOSE AND DIRECTION STATEMENT

❑ PRODUCT SIMULATION

❑ DEFINITION VIDEO

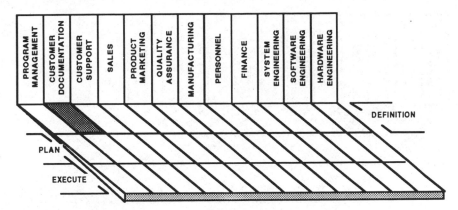

PHASE: DEFINITION   **TEAM**: CUSTOMER DOCUMENTATION

OUTPUT:

- ❑ SKELETON USER MANUALS
- ❑ CONCISE WRITTEN DESCRIPTION OF OTHER OUTPUT ITEMS
  - ◊ CUSTOMER TELEPHONE HOT-LINE
- ❑ ESTIMATE OF EXECUTION COST
- ❑ ESTIMATE OF COMPLEXITY
- ❑ ESTIMATE OF FULL STAFF NEEDS
- ❑ $TTM_{cust\ doc}$ ESTIMATE

SUGGESTIONS/REQUIREMENTS:

- ❑ $TTM_{cust\ doc}$ MODEL
- ❑ HISTORICAL PRODUCTIVITY DATA
- ❑ EXAMPLES OF PREVIOUS OUTPUT ITEMS

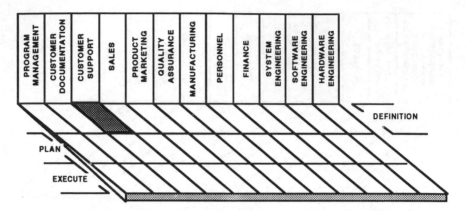

PHASE: DEFINITION   **TEAM**: CUSTOMER SUPPORT

OUTPUT:

- ❏ SERVICE AND REPAIR STRATEGY DEFINED
- ❏ SKELETON SUPPORT MANUALS
- ❏ CONCISE WRITTEN DESCRIPTION OF OTHER OUTPUT ITEMS
    - ◊ SERVICE PERSONNEL HOT-LINE SYSTEM
- ❏ ESTIMATE OF EXECUTION COST
- ❏ ESTIMATE OF COMPLEXITY
- ❏ ESTIMATE OF FULL STAFF NEEDS
- ❏ $TTM_{cust\ suppt}$ ESTIMATE

SUGGESTIONS/REQUIREMENTS:

- ❏ $TTM_{cust\ suppt}$ MODEL
- ❏ HISTORICAL PRODUCTIVITY DATA
- ❏ EXAMPLES OF PREVIOUS OUTPUT ITEMS

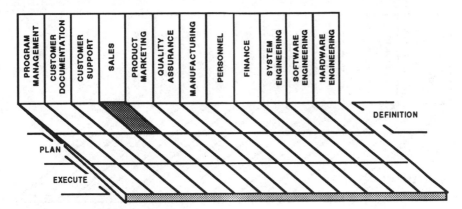

PHASE: DEFINITION   **TEAM**: SALES

OUTPUT:

❑ SALES FORECAST

❑ CONCISE WRITTEN DESCRIPTION OF OTHER OUTPUT ITEMS

  ◊ SALES PERSON TRAINING MATERIALS
  ◊ SALES PERSON PROMOTION EVENTS
  ◊ ORDER PROCESSING SYSTEM PROGRAMMING

❑ ESTIMATE OF EXECUTION COST

❑ ESTIMATE OF COMPLEXITY

❑ ESTIMATE OF FULL STAFF NEEDS

❑ TTM_{sales} ESTIMATE

SUGGESTIONS/REQUIREMENTS:

❑ TTM_{sales} MODEL

❑ HISTORICAL PRODUCTIVITY DATA

❑ EXAMPLES OF PREVIOUS OUTPUT ITEMS

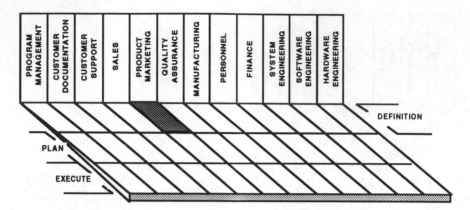

PHASE: DEFINITION   **TEAM**: PRODUCT MARKETING

OUTPUT:

❏ DRAFT DATA SHEETS

◊ FUNCTIONALITY   ◊ LOCALIZABILITY
◊ USABILITY   ◊ RELIABILITY
◊ PRICE   ◊ PERFORMANCE
◊ FORM AND FIT   ◊ SUPPORTABILITY

❏ COMPATIBILITY/UPGRADE STRATEGY DOCUMENT

❏ CONCISE WRITTEN DESCRIPTION OF OTHER DELIVERABLE ITEMS

◊ PRODUCT PROMOTIONS   ◊ PRODUCT ADVERTISING
◊ PROMOTION LITERATURE   ◊ MEDIA EVENTS

❏ ESTIMATE OF EXECUTION COST

❏ ESTIMATE OF COMPLEXITY

❏ ESTIMATE OF FULL STAFF NEEDS

❏ $TTM_{prod\ mktg}$ ESTIMATE

SUGGESTIONS/REQUIREMENTS:

❏ $TTM_{prod\ mktg}$ MODEL

❏ HISTORICAL PRODUCTIVITY DATA

❏ EXAMPLES OF PREVIOUS OUTPUT ITEMS

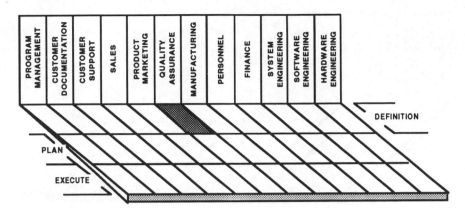

PHASE: DEFINITION   **TEAM**: QUALITY ASSURANCE

OUTPUT:

- ❑ PRODUCT TEST/QUALITY PLAN SKELETON

    - ◊ REGULATORY TESTS
    - ◊ ENVIRONMENTAL TESTS   ◊ RELIABILITY TESTS
    - ◊ DEFECT DENSITY TESTS   ◊ SAFETY TESTS

- ❑ CONCISE WRITTEN DESCRIPTION OF OTHER DELIVERABLE ITEMS

    - ◊ DEFINITION OF QUALITY ATTRIBUTES
    - ◊ FAILURE MODE IDENTIFICATION SHEETS

- ❑ ESTIMATE OF COMPLEXITY

- ❑ ESTIMATE OF EXECUTION COST

- ❑ ESTIMATE OF FULL STAFF NEEDS

- ❑ TTM_{qa} ESTIMATE

SUGGESTIONS/REQUIREMENTS:

- ❑ TTM_{qa} MODEL

- ❑ HISTORICAL PRODUCTIVITY DATA

- ❑ EXAMPLES OF PREVIOUS OUTPUT ITEMS

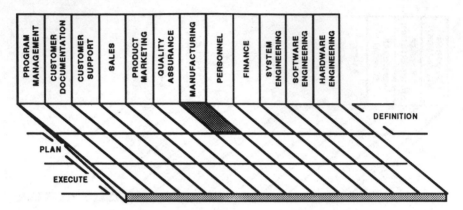

PHASE: DEFINITION   **TEAM**: MANUFACTURING

OUTPUT:

❑ DESIGN FOR MANUFACTURING PROCESS DOCUMENTS

    ◊ INFORMATION REQMTS   ◊ CYCLE TIMES
    ◊ WAGE & OVERHEAD RATES   ◊ VOLUME DEPENDENCIES

❑ PROCESS COST MODELS

    ◊ SOFTWARE   ◊ MATERIAL ACQUISITION
    ◊ PRINTED CIRCUITS   ◊ MATERIAL HANDLING
    ◊ PRODUCT ASSEMBLY   ◊ MATERIAL DISTRIBUTION
    ◊ PRODUCT TEST   ◊ MATERIAL FABRICATION
    ◊ INTEGRATED CIRCUITS   ◊ HYBRIDS

❑ CONCISE WRITTEN DESCRIPTION OF OTHER DELIVERABLE ITEMS

    ◊ ENGINEERING SERVICES   ◊ MATERIAL SERVICES

❑ TTM_{mfg} ESTIMATE

❑ ESTIMATE OF FULL STAFF NEEDS

❑ ESTIMATE OF EXECUTION COST

❑ ESTIMATE OF COMPLEXITY

SUGGESTIONS/REQUIREMENTS:

❑ TTM_{mfg} MODEL

❑ HISTORICAL PRODUCTIVITY DATA

❑ EXAMPLES OF PREVIOUS OUTPUT ITEMS

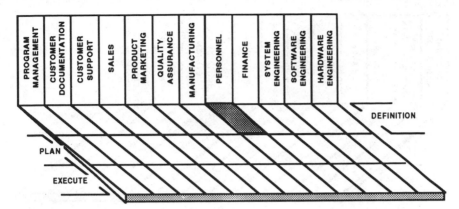

PHASE: DEFINITION   **TEAM**: PERSONNEL

OUTPUT:

- ❏ NEW HIRE REQUISITIONS

- ❏ RECRUITMENT PLAN SKELETON

- ❏ CONCISE WRITTEN DESCRIPTION OF OTHER DELIVERABLE ITEMS

    - ◊ WAGE AND SALARY SURVEYS
    - ◊ PROFESSIONAL HIRE ESTIMATES
    - ◊ AFFIRMATIVE ACTION GOALS
    - ◊ COLLEGE HIRE ESTIMATES
    - ◊ TRAINING PROGRAMS

- ❏ ESTIMATE OF COMPLEXITY

- ❏ ESTIMATE OF EXECUTION COST

- ❏ ESTIMATE OF FULL STAFF NEEDS

- ❏ TTM_{pers} ESTIMATE

SUGGESTIONS/REQUIREMENTS:

- ❏ TTM_{pers} MODEL

- ❏ HISTORICAL PRODUCTIVITY DATA

    - ◊ INTERVIEWS PER HIRE
    - ◊ CYCLE TIME/HIRE

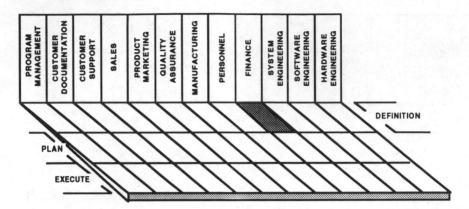

PHASE: DEFINITION   **TEAM**: FINANCE

OUTPUT:

- ❏ PRELIMINARY ESTIMATE OF PRODUCT COST

- ❏ FINANCIAL SENSITIVITY ANALYSIS SPREADSHEET

    ◊ IRR vs. TTM   ◊ IRR vs. EXECUTION $
    ◊ BREAK EVEN TIME vs. TTM   ◊ BET vs. EXECUTION $

- ❏ CONCISE WRITTEN DESCRIPTION OF OTHER DELIVERABLE ITEMS

    ◊ PROFITABILITY ANALYSIS
    ◊ FINANCIAL TRACKING SYSTEM INFORMATION

- ❏ ESTIMATE OF EXECUTION COST

- ❏ ESTIMATE OF COMPLEXITY

- ❏ ESTIMATE OF FULL STAFF NEEDS

- ❏ TTM_{fin} ESTIMATE

SUGGESTIONS/REQUIREMENTS:

- ❏ TTM_{fin} MODEL

- ❏ RETURN ON INVESTMENT MODEL DEFINED,
  EXPLAINED AND UNDERSTOOD

- ❏ HISTORICAL PRODUCTIVITY DATA

- ❏ CONCISE DESCRIPTION OF THE ACCOUNTING SYSTEM
  AVAILABLE FOR REVIEW

- ❏ CONCISE DESCRIPTION OF FINANCIAL MODELS
  AVAILABLE FOR REVIEW

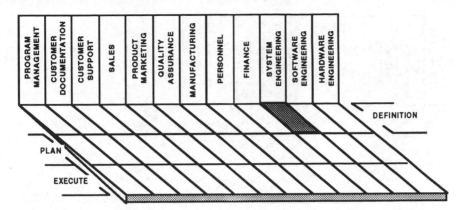

PHASE: DEFINITION   **TEAM:** SYSTEM ENGINEERING

OUTPUT:

❏ SYSTEM SPECIFICATION

◊ HARDWARE ELEMENTS   ◊ SOFTWARE ELEMENTS
◊ INTERCONNECT DFN   ◊ HUMAN INTERFACE DFN
◊ MACHINE INTERFACE DFN   ◊ FUNCTIONALITY
◊ LOCALIZABILITY   ◊ USABILITY
◊ RELIABILITY   ◊ COST
◊ SUPPORTABILITY   ◊ PERFORMANCE
◊ FORM   ◊ FIT

❏ SYSTEM BLOCK DIAGRAM

❏ PROTOTYPE PERFORMANCE DATA

❏ ESTIMATE OF EXECUTION COST

❏ ESTIMATE OF COMPLEXITY

❏ ESTIMATE OF FULL STAFF NEEDS

❏ TTM_{sys} ESTIMATE

SUGGESTIONS/REQUIREMENTS:

❏ TTM_{sys} MODEL

❏ HISTORICAL PRODUCTIVITY DATA

❏ EXAMPLES OF PREVIOUS OUTPUT ITEMS

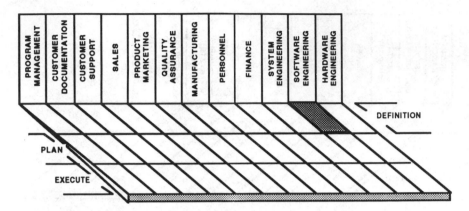

PHASE: DEFINITION   **TEAM**: SOFTWARE ENGINEERING

OUTPUT:

- ❏ PROTOTYPE SPECIFICATIONS

  - ◊ MODULE SPECIFICATIONS
  - ◊ KEY MODULE PERFORMANCE VERIFICATION
  - ◊ SIMULATION DATA

- ❏ PROTOTYPE DATA SHEET SIGN-OFF

  - ◊ FUNCTIONALITY   ◊ LOCALIZABILITY
  - ◊ RELIABILITY   ◊ USABILITY
  - ◊ COST   ◊ PERFORMANCE
  - ◊ SUPPORTABILITY   ◊ STYLE/DESIGN

- ❏ PROTOTYPE ERS WITH BLOCK DIAGRAM

- ❏ ESTIMATE OF COMPLEXITY

- ❏ ESTIMATE OF FULL STAFF NEEDS

- ❏ $TTM_{s/w}$ ESTIMATE

- ❏ TOOL NEEDS IDENTIFIED

- ❏ TEST PLAN SKELETON

SUGGESTIONS/REQUIREMENTS:

- ❏ $TTM_{s/w}$ MODEL

- ❏ HISTORICAL PRODUCTIVITY DATA

- ❏ EXAMPLES OF PREVIOUS DELIVERABLE ITEMS

- ❏ TTM "BEST PRACTICE" STORIES

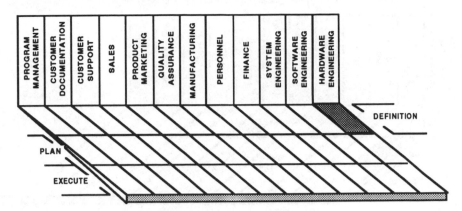

PHASE: DEFINITION   **TEAM**: HARDWARE ENGINEERING

OUTPUT:

- ❏ PROTOTYPE SPECIFICATIONS

    - ◊ MODULE SPECIFICATIONS
    - ◊ KEY MODULE PERFORMANCE VERIFICATION
    - ◊ HARDWARE SIMULATION DATA

- ❏ PROTOTYPE DATA SHEET SIGN-OFF

    - ◊ FUNCTIONALITY   ◊ LOCALIZABILITY
    - ◊ RELIABILITY   ◊ USABILITY
    - ◊ COST   ◊ PERFORMANCE
    - ◊ FORM AND FIT   ◊ SUPPORTABILITY

- ❏ PROTOTYPE ERS WITH BLOCK DIAGRAM

- ❏ TOOL NEEDS IDENTIFIED

- ❏ $TTM_{h/w}$ ESTIMATE

- ❏ TEST PLAN SKELETON

- ❏ ESTIMATE OF COMPLEXITY

- ❏ ESTIMATE OF FULL STAFF NEEDS

SUGGESTIONS/REQUIREMENTS:

- ❏ $TTM_{h/w}$ MODEL

- ❏ HISTORICAL PRODUCTIVITY DATA

- ❏ EXAMPLES OF PREVIOUS OUTPUT ITEMS

- ❏ TTM "BEST PRACTICE" STORIES

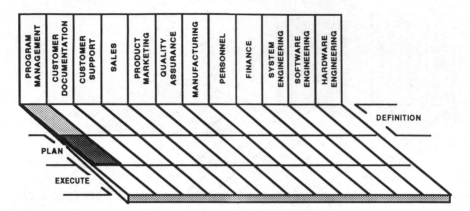

PHASE: PLANNING   **TEAM**: PROGRAM MANAGEMENT

OUTPUT:

❑   PRODUCT DEVELOPMENT PLAN SUMMARY

◊ CUST. DOCUMENTATION   ◊ CUSTOMER SUPPORT
◊ SALES   ◊ PRODUCT MARKETING
◊ QUALITY ASSURANCE   ◊ MANUFACTURING
◊ PERSONNEL   ◊ FINANCE
◊ SYSTEM ENGINEERING   ◊ SOFTWARE ENGINEERING
◊ HARDWARE ENGINEERING   ◊ PROJECT/PROGRAM MGMT

❑   INTEGRATED PRODUCT PERT NETWORK

❑   PRODUCT KMET CHART SHOWING BASELINE PLAN

❑ TTM_{pgm} TARGET

❑   CLOSURE WITH TTM_{pgm} ESTIMATE AND DEFINITION
       ADJUSTMENT

❑   PRESENTATION MATERIALS TO SECURE ORGANIZATION
       "GO AHEAD" AND RESOURCE SUPPORT TO ENTER THE
       EXECUTION PHASE

❑   ACTIVITY ASSIGNMENT REPORTS FOR THE TEAM

❑   FIRST ACTIVITY DETAIL REPORT

❑   PLAN TRACKING AND UPDATE METHODS ESTABLISHED

SUGGESTIONS/REQUIREMENTS:

- ❏ SOFTWARE TOOLS SELECTED

- ❏ EXAMPLE OF A PRODUCT DEVELOPMENT PLAN SUMMARY

- ❏ EXAMPLE OF AN INTEGRATED PRODUCT PERT NETWORK

- ❏ A "GENERIC" PROJECT/PROGRAM LEVEL NETWORK MODEL

- ❏ HISTORY OF TASK DURATION DATA BY TEAM

- ❏ PLAN SYNTHESIS TOOL KIT

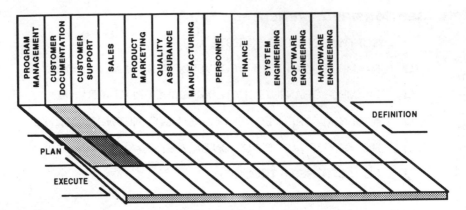

PHASE: PLANNING   **TEAM**: CUSTOMER DOCUMENTATION

OUTPUT:

❏   WORK BREAKDOWN BY DELIVERABLE ITEM

◊ VOLUME I   ◊ DISK I
    CHAPTER I
    CHAPTER II
    CHAPTER III, ETC.
◊ VOLUME II   ◊ DISK II
    CHAPTER I
    CHAPTER II
    CHAPTER III, ETC

❏   CUSTOMER DOCUMENTATION PERT NETWORK
    SHOWING THE FLOW OF OUTPUT OVER TIME

❏   CUSTOMER DOCUMENTATION KMET CHART

❏ $TTM_{cust\ doc}$ TARGET

❏   CLOSURE WITH $TTM_{cust\ doc}$ ESTIMATE AND DEFINITION
    ADJUSTMENT

❏   RESOURCES AVAILABLE FOR ASSIGNMENT

❏   RESOURCES NEEDED AS A FUNCTION OF TIME

❏   EXECUTION PHASE COST TARGET

❏   EXECUTION PHASE COMPLEXITY TARGET

SUGGESTIONS/REQUIREMENTS:

- ❏ EXAMPLE OF A WORK BREAKDOWN STRUCTURE

- ❏ EXAMPLE OF A CUSTOMER DOCUMENTATION PERT NETWORK

- ❏ A "GENERIC" CUSTOMER DOCUMENTATION NETWORK MODEL

- ❏ HISTORY OF CUSTOMER DOCUMENTATION TASK DURATION DATA

- ❏ PLAN SYNTHESIS TOOL KIT

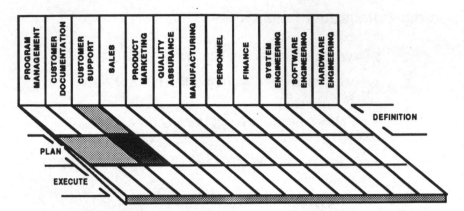

PHASE: PLANNING   **TEAM**: CUSTOMER SUPPORT

OUTPUT:

❑   WORK BREAKDOWN BY DELIVERABLE ITEM

◊ VOLUME I   ◊ DISK I
CHAPTER I
CHAPTER II
CHAPTER III, ETC.
◊ VOLUME II   ◊ DISK II
CHAPTER I
CHAPTER II
CHAPTER III, ETC.

❑   CUSTOMER SUPPORT PERT NETWORK
SHOWING THE FLOW OF OUTPUT OVER TIME

❑   CUSTOMER SUPPORT KMET CHART

❑ $TTM_{cust\ suppt}$ TARGET

❑   CLOSURE WITH $TTM_{cust\ suppt}$ ESTIMATE AND DEFINITION
ADJUSTMENT

❑   RESOURCES AVAILABLE FOR ASSIGNMENT

❑   RESOURCES NEEDED AS A FUNCTION OF TIME

❑   EXECUTION PHASE COST TARGET

❑   EXECUTION PHASE COMPLEXITY TARGET

SUGGESTIONS/REQUIREMENTS:

- ❑   EXAMPLE OF A WORK BREAKDOWN STRUCTURE
- ❑   EXAMPLE OF A CUSTOMER SUPPORT PERT NETWORK
- ❑   A "GENERIC" CUSTOMER SUPPORT NETWORK MODEL
- ❑   HISTORY OF CUSTOMER SUPPORT TASK DURATION DATA
- ❑   PLAN SYNTHESIS TOOL KIT

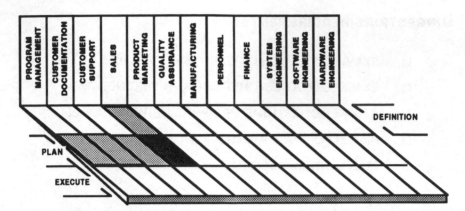

PHASE: PLANNING   **TEAM**: SALES

OUTPUT:

❑   WORK BREAKDOWN BY DELIVERABLE ITEM

  ◊ SALESPERSON TRAINING MANUAL
  ◊ SALESPERSON TRAINING VIDEO
  ◊ SALESPERSON MOTIVATION EVENTS
  ◊ ORDER PROCESSING SYSTEM PROGRAMMING

❑   SALES PERT NETWORK SHOWING THE FLOW OF
    OUTPUT AS A FUNCTION OF TIME

❑   SALES KMET CHART

❑ TTM_{sales} TARGET

❑   CLOSURE WITH TTM_{sales} ESTIMATE AND DEFINITION
    ADJUSTMENT

❑   RESOURCES AVAILABLE FOR ASSIGNMENT

❑   RESOURCES NEEDED AS A FUNCTION OF TIME

❑   EXECUTION PHASE COST TARGET

❑   EXECUTION PHASE COMPLEXITY TARGET

SUGGESTIONS/REQUIREMENTS:

- ❑   EXAMPLE OF A WORK BREAKDOWN STRUCTURE

- ❑   EXAMPLE OF A SALES PERT NETWORK

- ❑   A "GENERIC" SALES NETWORK MODEL

- ❑   HISTORY OF SALES TASK DURATION DATA

- ❑   PLAN SYNTHESIS TOOL KIT

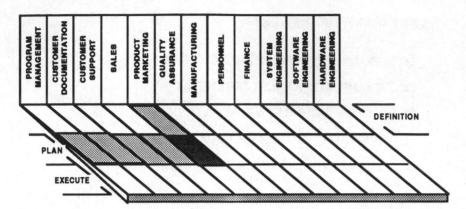

PHASE: PLANNING   **TEAM**: PRODUCT MARKETING

OUTPUT:

❑   WORK BREAKDOWN BY DELIVERABLE ITEM

   ◊ PRODUCT DATA SHEETS   ◊ PROMOTION LITERATURE
   ◊ MEDIA EVENTS   ◊ PRODUCT ADVERTISING
   ◊ COMPETITIVE ANALYSIS   ◊ COMPETITIVE POSITIONING
   ◊ TECHNICAL ARTICLES   ◊ SUCCESS STORIES

❑   MARKETING PERT NETWORK SHOWING THE FLOW
     OF OUTPUT OVER TIME

❑   MARKETING KMET CHART

❑ $TTM_{prod\ mktg}$ TARGET

❑   CLOSURE WITH $TTM_{prod\ mktg}$ ESTIMATE AND DEFINITION
     ADJUSTMENT

❑   RESOURCES AVAILABLE FOR ASSIGNMENT

❑   RESOURCES NEEDED AS A FUNCTION OF TIME

❑   EXECUTION PHASE COST TARGET

❑   EXECUTION PHASE COMPLEXITY TARGET

SUGGESTIONS/REQUIREMENTS:

- ❑ EXAMPLE OF A WORK BREAKDOWN STRUCTURE

- ❑ EXAMPLE OF A PRODUCT MARKETING PERT NETWORK

- ❑ A "GENERIC" PRODUCT MARKETING NETWORK MODEL

- ❑ HISTORY OF PRODUCT MARKETING TASK DURATION DATA

- ❑ PLAN SYNTHESIS TOOL KIT

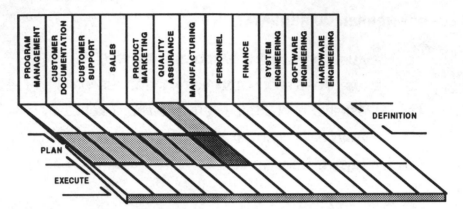

PHASE: PLANNING   **TEAM**: QUALITY ASSURANCE

OUTPUT:

❑   WORK BREAKDOWN BY DELIVERABLE ITEM

◊ ENVIRONMENTAL TEST   ◊ RELIABILITY TEST
◊ REGULATORY AGENCY TESTS   ◊ SAFETY TEST
◊ DEFECT DENSITY TESTS   ◊ QUALITY AUDITS
◊ FAILURE RATE FORECASTS   ◊ DEFECT ANALYSIS

❑   QUALITY ASSURANCE PERT NETWORK SHOWING
    THE FLOW OF OUTPUT OVER TIME

❑   QUALITY ASSURANCE KMET CHART

❑ TTM_{qa} TARGET

❑   CLOSURE WITH TTM_{qa} ESTIMATE AND DEFINITION
    ADJUSTMENT

❑   RESOURCES AVAILABLE FOR ASSIGNMENT

❑   RESOURCES NEEDED AS A FUNCTION OF TIME

❑   EXECUTION PHASE COST TARGET

❑   EXECUTION PHASE COMPLEXITY TARGET

SUGGESTIONS/REQUIREMENTS:

- ❏ EXAMPLE OF A WORK BREAKDOWN STRUCTURE

- ❏ EXAMPLE OF A QUALITY ASSURANCE PERT NETWORK

- ❏ A "GENERIC" QUALITY ASSURANCE NETWORK MODEL

- ❏ HISTORY OF QUALITY ASSURANCE TASK DURATION DATA

- ❏ PLAN SYNTHESIS TOOL KIT

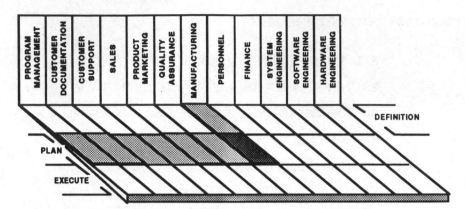

PHASE: PLANNING   **TEAM:** MANUFACTURING

OUTPUT:

❑ **WORK BREAKDOWN BY DELIVERABLE ITEM**

◊ SOFTWARE MFG   ◊ MATERIAL ACQUISITION
◊ PRINTED CIRCUIT BOARDS ◊ MATERIAL HANDLING
◊ PRODUCT ASSEMBLY   ◊ MATERIAL DISTRIBUTION
◊ PRODUCT TEST   ◊ MATERIAL FABRICATION
◊ INTEGRATED CIRCUITS   ◊ HYBRIDS
◊ MIS DATA INPUT   ◊ MATL SYS DATA INPUT

❑ **MANUFACTURING PERT NETWORK SHOWING FLOW OF OUTPUT OVER TIME**

❑ **MANUFACTURING KMET CHART**

❑ **TTM$_{mfg}$ TARGET**

❑ **CLOSURE WITH TTM $_{mfg}$ ESTIMATE AND DEFINITION ADJUSTMENT**

❑ **RESOURCES AVAILABLE FOR ASSIGNMENT**

❑ **RESOURCES NEEDED AS A FUNCTION OF TIME**

❑ **EXECUTION PHASE COST TARGET**

❑ **EXECUTION PHASE COMPLEXITY TARGET**

SUGGESTIONS/REQUIREMENTS:

❏   EXAMPLE OF A WORK BREAKDOWN STRUCTURE

❏   EXAMPLE OF A MANUFACTURING PERT NETWORK

❏   A "GENERIC" MANUFACTURING NETWORK MODEL

❏   HISTORY OF MANUFACTURING TASK DURATION DATA

◊ MINUTES/UNIT TO ASSEMBLE
◊ MINUTES/UNIT TO TEST
◊ PER UNIT CYCLE TIME

❏   PLAN SYNTHESIS TOOL KIT

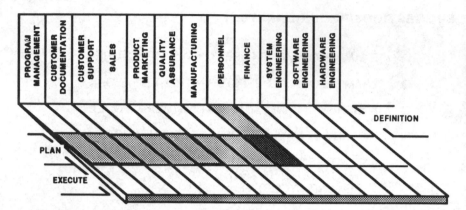

PHASE: PLANNING   **TEAM**: PERSONNEL

OUTPUT:

- ❏ WORK BREAKDOWN BY DELIVERABLE ITEM

    - ◊ RECRUITMENT
    - ◊ TRAINING
    - ◊ EVALUATION
    - ◊ COUNSELING
    - ◊ TERMINATION

- ❏ PERSONNEL PERT NETWORK SHOWING THE FLOW OF OUTPUT OVER TIME

- ❏ PERSONNEL KMET CHART

- ❏ TTM_{pers} TARGET

- ❏ CLOSURE WITH TTM_{pers} ESTIMATE AND DEFINITION ADJUSTMENT

- ❏ RESOURCES AVAILABLE FOR ASSIGNMENT

- ❏ RESOURCES NEEDED AS A FUNCTION OF TIME

- ❏ EXECUTION PHASE COST TARGET

- ❏ EXECUTION PHASE COMPLEXITY TARGET

SUGGESTIONS/REQUIREMENTS:

❑   EXAMPLE OF A WORK BREAKDOWN STRUCTURE

❑   EXAMPLE OF A PERSONNEL PERT NETWORK

❑   A "GENERIC" PERSONNEL NETWORK MODEL

❑   HISTORY OF PERSONNEL TASK DURATION DATA

◊ TIME TO HIRE EXPERIENCE   ◊ TIME TO HIRE COLLEGE
◊ TIME TO HIRE MGMT   ◊ TIME TO TRAIN
◊ EXPECTED TURNOVER RATE   ◊ TIME TO RETRAIN

❑   PLAN SYNTHESIS TOOL KIT

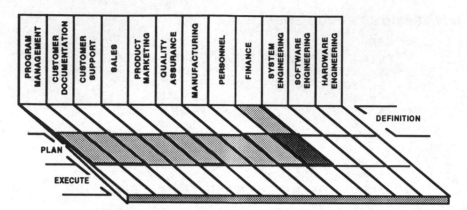

PHASE: PLANNING   **TEAM:** FINANCE

OUTPUT:

❏   WORK BREAKDOWN BY DELIVERABLE ITEM

◊   ENGINEERING $ vs. TIME   ◊   MFG $ vs. TIME
◊   MARKETING $ vs. TIME   ◊   OTHER'S $ vs. TIME
◊   PRICING MODEL UPDATES   ◊   IRR vs. TTM UPDATE
◊   FINANCE SYS DATA INPUT   ◊   SYS DATA OUTPUT

❏   FINANCE PERT NETWORK SHOWING THE FLOW OF
      OUTPUT OVER TIME

❏   FINANCE KMET CHART

❏ TTM_{fin} TARGET

❏   CLOSURE WITH TTM_{fin} ESTIMATE AND DEFINITION
      ADJUSTMENT

❏   RESOURCES AVAILABLE FOR ASSIGNMENT

❏   RESOURCES NEEDED AS A FUNCTION OF TIME

❏   EXECUTION PHASE COST TARGET

❏   EXECUTION PHASE COMPLEXITY TARGET

SUGGESTIONS/REQUIREMENTS:

❑   EXAMPLE OF A WORK BREAKDOWN STRUCTURE

❑   EXAMPLE OF A FINANCE PERT NETWORK

❑   A "GENERIC" FINANCE NETWORK MODEL

❑   HISTORY OF FINANCE TASK DURATION DATA

❑   PLAN SYNTHESIS TOOL KIT

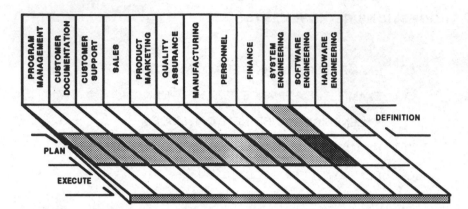

PHASE: PLANNING   **TEAM:** SYSTEM ENGINEERING

OUTPUT:

    ❏   WORK BREAKDOWN BY DELIVERABLE ITEM

         ◊ H/W SPEC CONTROL   ◊ S/W SPEC CONTROL
         ◊ SYSTEM SPEC CONTROL ◊ PRE-MFG UNIT BUILDS
         ◊ PRE-MFG SYSTEM BUILDS ◊ SYSTEM INTEGRATION
         ◊ SYSTEM EVALUATION   ◊ SYSTEM REWORK

    ❏   SYSTEM ENGINEERING PERT NETWORK SHOWING THE FLOW OF OUTPUT OVER TIME

    ❏   SYSTEM ENGINEERING KMET CHART

    ❏ TTM_{sys} TARGET

    ❏   CLOSURE WITH TTM_{sys} ESTIMATE AND DEFINITION ADJUSTMENT

    ❏   RESOURCES AVAILABLE FOR ASSIGNMENT

    ❏   RESOURCES NEEDED AS A FUNCTION OF TIME

    ❏   EXECUTION PHASE COST TARGET

    ❏   EXECUTION PHASE COMPLEXITY TARGET

SUGGESTIONS/REQUIREMENTS:

❑   EXAMPLE OF A WORK BREAKDOWN STRUCTURE

❑   EXAMPLE OF A SYSTEM ENGINEERING PERT NETWORK

❑   A "GENERIC" SYSTEM ENGINEERING NETWORK MODEL

❑   HISTORY OF SYSTEM ENGINEERING TASK DURATION DATA

◊ INTEGRATION TIME PER UNIT   ◊ FACTORY TEST TIME
◊ FIELD TEST TIME   ◊ REWORK TIME

❑   PLAN SYNTHESIS TOOL KIT

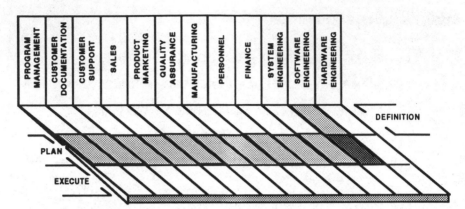

PHASE: PLANNING   **TEAM**: SOFTWARE ENGINEERING

OUTPUT:

❑   WORK BREAKDOWN BY DELIVERABLE ITEM

◊ DESIGNED MODULES ◊ CODED MODULES
◊ TESTED MODULES ◊ MODULE INTEGRATION
◊ SOFTWARE TEST ◊ SOFTWARE REWORK
◊ SYSTEM REWORK ◊ SYSTEM TEST SUPPORT
◊ INTEGRATION SUPPORT ◊ SYSTEM CONSULTATION

❑   SOFTWARE ENGINEERING PERT NETWORK SHOWING THE
    FLOW OF OUTPUT OVER TIME

❑   SOFTWARE KMET CHART

❑ $TTM_{s/w}$ TARGET

❑   CLOSURE WITH $TTM_{s/w}$ ESTIMATE AND DEFINITION
    ADJUSTMENT

❑   RESOURCES AVAILABLE FOR ASSIGNMENT

❑   RESOURCES NEEDED AS A FUNCTION OF TIME

❑   EXECUTION PHASE COST TARGET

❑   EXECUTION PHASE COMPLEXITY TARGET

SUGGESTIONS/REQUIREMENTS:

- ❏   EXAMPLE OF A WORK BREAKDOWN STRUCTURE

- ❏   EXAMPLE OF A SOFTWARE ENGINEERING PERT NETWORK

- ❏   A "GENERIC" SOFTWARE ENGINEERING NETWORK MODEL

- ❏   HISTORY OF SOFTWARE ENGINEERING TASK DURATION
     DATA

     ◊ NON REAL-TIME MODULES
     ◊ REAL-TIME MODULES

- ❏   PLAN SYNTHESIS TOOL KIT

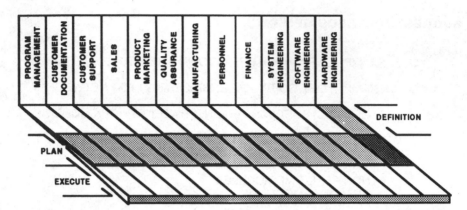

PHASE: PLANNING   **TEAM**: HARDWARE ENGINEERING

OUTPUT:

❑   WORK BREAKDOWN BY DELIVERABLE ITEM

◊ PC DESIGNS   ◊ IC DESIGNS
◊ HYBRID DESIGNS   ◊ MECHANICAL DESIGNS
◊ UNIT (BOX) DESIGNS   ◊ DESIGN DOCUMENTS
◊ PRE-MFG H/W BUILDS   ◊ MODULE TEST
◊ UNIT TEST   ◊ REDESIGN
◊ RETEST

❑   HARDWARE ENGINEERING PERT NETWORK SHOWING THE
     FLOW OF OUTPUT OVER TIME

❑   HARDWARE ENGINEERING KMET CHART

❑ $TTM_{h/w}$ TARGET

❑   CLOSURE WITH $TTM_{h/w}$ ESTIMATE AND DEFINITION
     ADJUSTMENT

❑   RESOURCES AVAILABLE FOR ASSIGNMENT

❑   RESOURCES NEEDED AS A FUNCTION OF TIME

❑   EXECUTION PHASE COST TARGET

❑   EXECUTION PHASE COMPLEXITY TARGET

SUGGESTIONS/REQUIREMENTS:

❏   EXAMPLE OF A WORK BREAKDOWN STRUCTURE

❏   EXAMPLE OF A HARDWARE ENGINEERING PERT NETWORK

❏   A "GENERIC" HARDWARE ENGINEERING NETWORK MODEL

❏   HISTORY OF HARDWARE ENGINEERING TASK DURATION
        DATA

◊ PC BOARD DESIGN   ◊ PC BOARD FAB
◊ PC BOARD LOAD   ◊ PC BOARD TEST

◊ IC LAYOUT   ◊ IC FAB
◊ IC PACKAGE   ◊ IC TEST

◊ MODULE ASSEMBLY   ◊ MODULE TEST

◊ UNIT ASSEMBLY   ◊ UNIT TEST

◊ REDESIGN   ◊ RETEST

❏   PLAN SYNTHESIS TOOL KIT

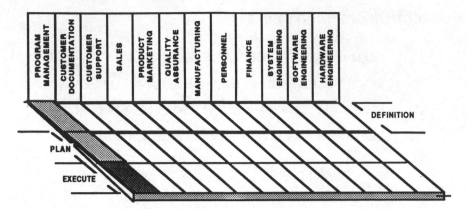

PHASE: EXECUTION   **TEAM**: PROGRAM MANAGEMENT

OUTPUT:

❑ PERIODIC REPORT ON NETWORK ACTIVITY

   ◊ TASKS ADDED   ◊ TASKS DELETED
   ◊ TASKS CHANGED

❑ PERIODIC COMPUTATION OF "SURVIVOR'S INDEX"

❑ PERIODIC ACTIVITY ASSIGNMENT REPORTS

❑ PERIODIC UPDATE OF PROJECT KMET CHART

❑ PERIODIC GENERATION OF THE SCHEDULE
   EXCEPTION REPORT

❑ TTM_{pgm} ACTUAL TRACKING

❑ PERIODIC PROGRESS/PROBLEM REVIEWS
   WITH FUNCTIONAL AND GENERAL MANAGEMENT

❑ EXECUTION PHASE COST TRACKING

❑ EXECUTION PHASE COMPLEXITY TRACKING

❑ EXECUTION PHASE STAFF TRACKING

SUGGESTIONS/REQUIREMENTS:

❑   WEEKLY MEETINGS WITH FUNCTIONAL TEAM LEADERS

❑   MILESTONE MEETINGS WITH THE ENTIRE
     PROJECT/PROGRAM TEAM

❑   REPORT GENERATION SOFTWARE FROM
     TIME TO MARKET ASSOCIATES

❑   ANALYSIS AND TRACKING SOFTWARE FROM
     TIME TO MARKET ASSOCIATES

❑   PROJECT/PROGRAM LIFE-CYCLE MODEL

❑   AUTOMATED, REAL-TIME TRACKING SYSTEM

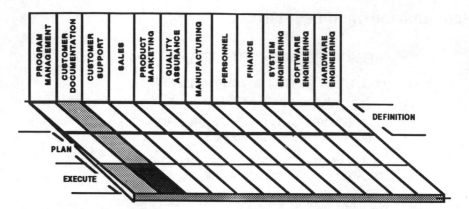

PHASE: EXECUTION   **TEAM**: CUSTOMER DOCUMENTATION

OUTPUT:

❑   INFORMATION TO PRODUCE CUSTOMER DOCUMENTATION
     DELIVERED TO THE MANUFACTURER

❑   PERIODIC REPORT ON NETWORK ACTIVITY TO
     PROJECT/PROGRAM MANAGER

     ◊ TASKS ADDED   ◊ TASKS DELETED
     ◊ TASKS CHANGED   ◊ LINKAGE REVISIONS
     ◊ RESOURCE REVISIONS   ◊ DURATION REVISIONS

❑   EXECUTION PHASE COST TRACKING

❑   EXECUTION PHASE COMPLEXITY TRACKING

❑   EXECUTION PHASE STAFF TRACKING

❑   TTM$_{cust\ doc}$ ACTUAL TRACKING

❑   CONCISE, PERIODIC, PROGRESS/PROBLEM REPORT
     TO THE PROJECT/PROGRAM MANAGER

SUGGESTIONS/REQUIREMENTS:

- ❑   ACCESS TO DESIGN ENGINEERS

- ❑   PRELIMINARY WRITE-UPS FROM THE DESIGN ENGINEERS

- ❑   ACCESS TO DESIGN DATA AND DOCUMENTATION

- ❑   ACCESS TO THE END PRODUCT

- ❑   TALENTED, MOTIVATED TEAM OF PEOPLE

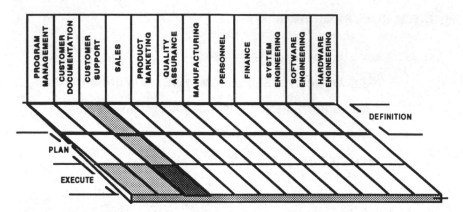

PHASE: EXECUTION   **TEAM**: CUSTOMER SUPPORT

OUTPUT:

❏ INFORMATION AND MATERIAL TO PROVIDE CUSTOMER
    SUPPORT DELIVERED TO THE FIELD SUPPORT NETWORK

 ◊ DOCUMENTATION
 ◊ CD ROM
 ◊ TELEPHONE SUPPORT SYSTEM
 ◊ INVENTORY OF SPARE PARTS ESTABLISHED

❏ PERIODIC REPORT ON NETWORK ACTIVITY TO
    PROJECT/PROGRAM MANAGER

 ◊ TASKS ADDED   ◊ TASKS DELETED
 ◊ TASKS CHANGED ◊ LINKAGE REVISIONS
 ◊ RESOURCE REVISIONS ◊ DURATION REVISIONS

❏ EXECUTION PHASE COST TRACKING

❏ EXECUTION PHASE COMPLEXITY TRACKING

❏ EXECUTION PHASE STAFF TRACKING

❏ $TTM_{cust\ suppt}$ ACTUAL TRACKING

❏ CONCISE, PERIODIC PROGRESS/PROBLEM REPORT
    TO THE PROJECT/PROGRAM MANAGER

SUGGESTIONS/REQUIREMENTS:

❑ ACCESS TO DESIGN ENGINEERS

❑ FAULT ISOLATION ALGORITHMS FROM DESIGN ENGINEERS

❑ PRELIMINARY PERFORMANCE VERIFICATION METHODS FROM DESIGN ENGINEERS

❑ ACCESS TO DESIGN DATA AND DOCUMENTATION

❑ ACCESS TO THE END PRODUCT

❑ TALENTED, MOTIVATED TEAM OF PEOPLE

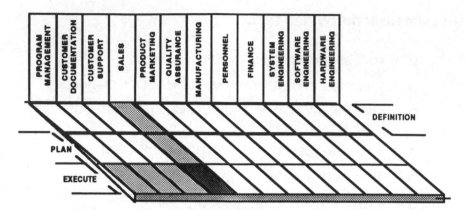

PHASE: EXECUTION   **TEAM**: SALES

OUTPUT:

- ❏ INFORMATION TO PRODUCE SALES SUPPORT MATERIALS DELIVERED TO THE MANUFACTURER

    - ◊ SALES TRAINING MANUALS   ◊ SALES TRAINING VIDEO

- ❏ SALESPERSON MOTIVATION EVENTS SCHEDULED AND STRUCTURED

- ❏ PERIODIC REPORT ON NETWORK ACTIVITY TO PROJECT/PROGRAM MANAGER

    - ◊ TASKS ADDED   ◊ TASKS DELETED
    - ◊ TASKS CHANGED   ◊ LINKAGE REVISIONS
    - ◊ RESOURCE REVISIONS   ◊ DURATION REVISIONS

- ❏ ORDER PROCESSING SYSTEM PROGRAMMED

- ❏ EXECUTION PHASE COST TRACKING

- ❏ EXECUTION PHASE COMPLEXITY TRACKING

- ❏ EXECUTION PHASE STAFF TRACKING

- ❏ TTM_{sales} ACTUAL TRACKING

- ❏ CONCISE, PERIODIC PROGRESS/PROBLEM REPORT TO THE PROJECT/PROGRAM MANAGER

SUGGESTIONS/REQUIREMENTS:

- ❑   DATABASE ON PRESENT CUSTOMERS BUYING PATTERNS

- ❑   SALES GENERATED BY COMPETITORS' PRODUCTS

- ❑   ACCESS TO DESIGN DATA AND DOCUMENTATION

- ❑   ACCESS TO THE END PRODUCT

- ❑   TALENTED, MOTIVATED TEAM OF PEOPLE

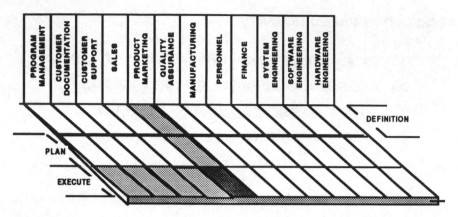

PHASE: EXECUTION   **TEAM**: PRODUCT MARKETING

OUTPUT:

- ❏ PRODUCT SALES PRICE ESTABLISHED

- ❏ INFORMATION TO PRODUCE PRODUCT MARKETING
  DOCUMENTATION DELIVERED TO THE MANUFACTURER

  - ◊ DATA SHEETS   ◊ PROMOTION LITERATURE
  - ◊ ADVERTISING MATERIALS ◊ MEDIA MATERIALS
  - ◊ COMPETITIVE ANALYSIS ◊ PRODUCT POSITIONING
  - ◊ APPLICATION NOTES   ◊ MAGAZINE ARTICLES

- ❏ PERIODIC REPORT ON NETWORK ACTIVITY TO
  PROJECT/PROGRAM MANAGER

  - ◊ TASKS ADDED   ◊ TASKS DELETED
  - ◊ TASKS CHANGED   ◊ LINKAGE REVISIONS
  - ◊ RESOURCE REVISIONS   ◊ DURATION REVISIONS

- ❏ EXECUTION PHASE COST TRACKING

- ❏ EXECUTION PHASE COMPLEXITY TRACKING

- ❏ EXECUTION PHASE STAFF TRACKING

- ❏ $TTM_{prod\ mktg}$ ACTUAL TRACKING

- ❏ CONCISE, PERIODIC PROGRESS/PROBLEM REPORT
  TO THE PROJECT/PROGRAM MANAGER

SUGGESTIONS/REQUIREMENTS:

- ❏ ACCESS TO DESIGN ENGINEERS

- ❏ INFORMAL INTERACTION WITH DESIGN ENGINEERS

- ❏ ACCESS TO AND INTERACTION WITH POTENTIAL CUSTOMERS

- ❏ ACCESS TO DESIGN DATA AND DOCUMENTATION

- ❏ ACCESS TO THE END PRODUCT

- ❏ TALENTED, MOTIVATED TEAM OF PEOPLE

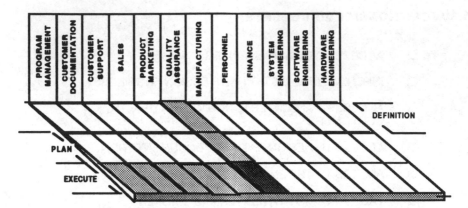

PHASE: EXECUTION   **TEAM**: QUALITY ASSURANCE

OUTPUT:

- ❏ ENVIRONMENTAL CERTIFICATION DELIVERED TO HARDWARE
     AND SYSTEMS ENGINEERING AND PLACED IN A PERMANENT
     QA FILE

- ❏ REGULATORY TEST RESULTS AND SAFETY TEST RESULTS
     PLACED IN PERMANENT QA FILE

- ❏ RELIABILITY TEST RESULTS DELIVERED TO ENGINEERING
     AND PROJECT/PROGRAM MANAGEMENT

- ❏ PERIODIC REPORT ON NETWORK ACTIVITY TO
     PROJECT/PROGRAM MANAGER

     - ◊ TASKS ADDED   ◊ TASKS DELETED
     - ◊ TASKS CHANGED   ◊ LINKAGE REVISIONS
     - ◊ RESOURCE REVISIONS   ◊ DURATION REVISIONS

- ❏ EXECUTION PHASE COST TRACKING

- ❏ EXECUTION PHASE COMPLEXITY TRACKING

- ❏ EXECUTION PHASE STAFF TRACKING

- ❏ TTM_{qa} ACTUAL TRACKING

- ❏ CONCISE, PERIODIC PROGRESS/PROBLEM REPORT
     TO THE PROJECT/PROGRAM MANAGER

SUGGESTIONS/REQUIREMENTS:

- ❏   ACCESS TO DESIGN ENGINEERS
- ❏   REFERENCE ENVIRONMENTAL TEST STANDARDS MANUAL
- ❏   REFERENCE SAFETY TEST STANDARDS MANUAL
- ❏   SOFTWARE QA STANDARDS MANUAL
- ❏   METHOD ESTABLISHED TO PREDICT ANNUAL FAILURE RATE
- ❏   ACCESS TO DESIGN DATA AND DOCUMENTATION
- ❏   ACCESS TO THE END PRODUCT
- ❏   TALENTED, MOTIVATED TEAM OF PEOPLE

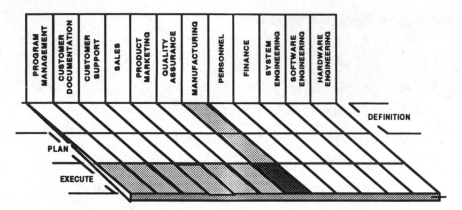

PHASE: EXECUTION   **TEAM**: MANUFACTURING

OUTPUT:

- ❏ "TOTAL" PRODUCT SUPPLIED TO THE CUSTOMER

    - ◊ CUSTOMER DOCUMENTATION
    - ◊ SOFTWARE
    - ◊ HARDWARE
    - ◊ SYSTEM

- ❏ PRODUCT DELIVERY GOALS ESTABLISHED

- ❏ MANUFACTURING COST INFORMATION SUPPLIED TO FINANCE

- ❏ MANUFACTURING SYSTEMS FULLY PROGRAMMED

- ❏ PERIODIC REPORT ON NETWORK ACTIVITY TO PROJECT/PROGRAM MANAGER

    - ◊ TASKS ADDED   ◊ TASKS DELETED
    - ◊ TASKS CHANGED   ◊ LINKAGE REVISIONS
    - ◊ RESOURCE REVISIONS   ◊ DURATION REVISIONS

- ❏ EXECUTION PHASE COST TRACKING

- ❏ EXECUTION PHASE COMPLEXITY TRACKING

- ❏ EXECUTION PHASE STAFF TRACKING

- ❏ TTM_{mfg} ACTUAL TRACKING

- ❏ CONCISE, PERIODIC PROGRESS/PROBLEM REPORT TO THE PROJECT/PROGRAM MANAGER

SUGGESTIONS/REQUIREMENTS:

- ❑ ACCESS TO DESIGN ENGINEERS AND PEOPLE IN OTHER AREAS

- ❑ MANUFACTURING SYSTEMS INPUT DATA REQUIREMENTS WELL SPECIFIED TO THOSE GENERATING INPUT DATA

- ❑ INFORMATION NEEDS BY MANUFACTURING PROCESS WELL SPECIFIED TO THOSE GENERATING PROCESS INPUT DATA

    - ◊ SOFTWARE
    - ◊ PRINTED CIRCUITS
    - ◊ HYBRID ASSEMBLY
    - ◊ MATL FABRICATION
    - ◊ PRODUCT ASSEMBLY
    - ◊ MATERIALS
    - ◊ INTEGRATED CIRCUITS
    - ◊ HYBRID TEST
    - ◊ SHIPPING
    - ◊ PRODUCT TEST

- ❑ ACCESS TO DESIGN DATA AND DOCUMENTATION

- ❑ ACCESS TO THE END PRODUCT

- ❑ TALENTED, MOTIVATED TEAM OF PEOPLE

- ❑ FINAL ASSEMBLY AND TEST METHODS MODELED AND UNDERSTOOD

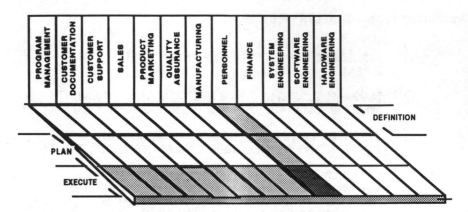

PHASE: EXECUTION   **TEAM**: PERSONNEL

OUTPUT:

❑ PROJECT/PROGRAM STAFFING NEEDS SATISFIED

❑ TRAINING PROGRAMS CONDUCTED

◊ ORIENTATION   ◊ INTERPERSONAL SKILLS
◊ TIME-TO-MARKET BASICS   ◊ PROJECT MGMT CLASSES
◊ TEAM MEMBER CLASSES   ◊ MANAGEMENT CLASSES

❑ PERIODIC REPORT ON NETWORK ACTIVITY TO
PROJECT/PROGRAM MANAGER

◊ TASKS ADDED   ◊ TASKS DELETED
◊ TASKS CHANGED   ◊ LINKAGE REVISIONS
◊ RESOURCE REVISIONS   ◊ DURATION REVISIONS

❑ EXECUTION PHASE COST TRACKING

❑ EXECUTION PHASE COMPLEXITY TRACKING

❑ EXECUTION PHASE STAFF TRACKING

❑ TTM_{pers} ACTUAL TRACKING

❑ CONCISE, PERIODIC, PROGRESS/PROBLEM REPORT
TO THE PROJECT/PROGRAM MANAGER

SUGGESTIONS/REQUIREMENTS:

- ❏ ACCESS TO FUNCTIONAL TEAM MANAGERS
- ❏ COLLEGE RECRUITING PROGRAM ESTABLISHED
- ❏ PROFESSIONAL RECRUITING PROGRAM ESTABLISHED
- ❏ INTERNAL TRAINING NEEDS DETERMINED
- ❏ TRAINING AND DEVELOPMENT COURSE LEADERS DETERMINED
- ❏ EMPLOYEE RETENTION PROGRAMS
- ❏ EMPLOYEE ASSISTANCE PROGRAMS
- ❏ TALENTED, MOTIVATED, TEAM OF PEOPLE

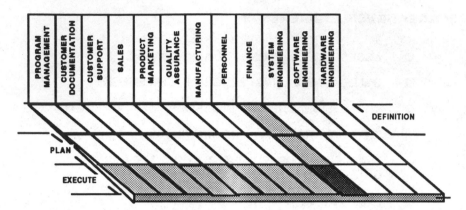

PHASE: EXECUTION   **TEAM**: FINANCE

OUTPUT:

- ❏ PERIODIC REPORTS ON THE PROJECT/PROGRAM
  EXPENDITURE OF FUNDS

    - ◊ ENGINEERING   ◊ MANUFACTURING
    - ◊ MARKETING   ◊ OTHERS

- ❏ PRODUCT PRICE ALTERNATIVES ANALYZED FOR PRODUCT
  MARKETING

- ❏ MECHANISM TO TRACK BREAK EVEN TIME ACTIVATED

- ❏ PERIODIC REPORT ON NETWORK ACTIVITY TO
  PROJECT/PROGRAM MANAGER

    - ◊ TASKS ADDED   ◊ TASKS DELETED
    - ◊ TASKS CHANGED   ◊ LINKAGE REVISIONS
    - ◊ RESOURCE REVISIONS   ◊ DURATION REVISIONS

- ❏ EXECUTION PHASE COST TRACKING

- ❏ EXECUTION PHASE COMPLEXITY TRACKING

- ❏ EXECUTION PHASE STAFF TRACKING

- ❏ TTM_{fin} ACTUAL TRACKING

- ❏ CONCISE, PERIODIC PROGRESS/PROBLEM REPORT
  TO THE PROJECT/PROGRAM MANAGER

SUGGESTIONS/REQUIREMENTS:

❑   AUTOMATED ACCESS TO EXPENSE DATA

❑   EXPENSE TARGETING PROCESS DRIVEN BY THE
     PROJECT/PROGRAM PLANNING PROCESS AND TTM

❑   APPROXIMATE, SIMPLE, SPREADSHEET FINANCIAL MODELS

❑   FINANCIAL TRACKING SYSTEM INFORMATION NEEDS
     WELL SPECIFIED

❑   TALENTED, MOTIVATED TEAM OF PEOPLE

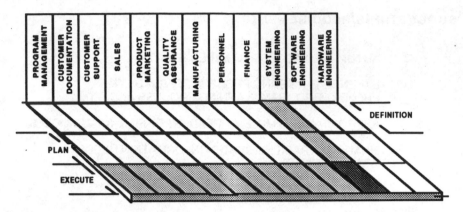

PHASE: EXECUTION   **TEAM**: SYSTEM ENGINEERING

OUTPUT:

❑ PRE-MANUFACTURING SYSTEMS AVAILABLE

❑ THOROUGH SYSTEM TEST

❑ ACTUAL SYSTEM PERFORMANCE CLOSURE WITH PREDICTIONS

❑ DESIGN INFORMATION SUPPLIED TO

◊ CUSTOMER DOCUMENTATION ◊ CUSTOMER SUPPORT
◊ PRODUCT MARKETING   ◊ QUALITY ASSURANCE
◊ MANUFACTURING   ◊ H/W ENGINEERING
◊ S/W ENGINEERING

❑ PERIODIC REPORT ON NETWORK ACTIVITY TO PROJECT/PROGRAM MANAGER

◊ TASKS ADDED   ◊ TASKS DELETED
◊ TASKS CHANGED   ◊ LINKAGE REVISIONS
◊ RESOURCE REVISIONS   ◊ DURATION REVISIONS

❑ EXECUTION PHASE COST TRACKING

❑ EXECUTION PHASE COMPLEXITY TRACKING

❑ EXECUTION PHASE STAFF TRACKING

❑ TTM_{sys} ACTUAL TRACKING

❑ CONCISE, PERIODIC PROGRESS/PROBLEM REPORT TO THE PROJECT/PROGRAM MANAGER

SUGGESTIONS/REQUIREMENTS:

- ❑   THOROUGH UNDERSTANDING OF THE INFORMATION NEEDED BY OTHERS

- ❑   DEFINED AND COMMUNICATED DOCUMENTATION STANDARDS

- ❑   DEFINED AND STRUCTURED SYSTEM EVALUATION METHODS

- ❑   ACCESSIBLE TO OTHER TEAM MEMBERS IN NEED OF INFORMATION CLARIFICATION

- ❑   ACCESS TO PRE-MANUFACTURING HARDWARE

- ❑   ACCESS TO PRE-MANUFACTURING SOFTWARE

- ❑   TALENTED, MOTIVATED TEAM OF PEOPLE

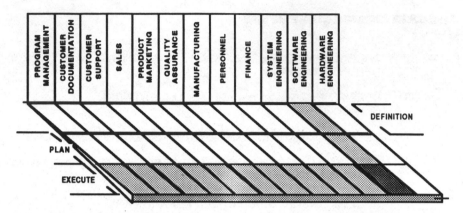

PHASE: EXECUTION   **TEAM**: SOFTWARE ENGINEERING

OUTPUT:

❑   PRE-MANUFACTURING SOFTWARE AVAILABLE

❑   THOROUGH SOFTWARE TEST

❑   ACTUAL SOFTWARE PERFORMANCE CLOSURE WITH
     PREDICTIONS

❑   DESIGN INFORMATION SUPPLIED TO

◊ CUSTOMER DOCUMENTATION   ◊ CUSTOMER SUPPORT
◊ PRODUCT MARKETING   ◊ QUALITY ASSURANCE
◊ MANUFACTURING   ◊ SYSTEM ENGINEERING
◊ HARDWARE ENGINEERING

❑   PERIODIC REPORT ON NETWORK ACTIVITY TO
     PROJECT/PROGRAM MANAGER

◊ TASKS ADDED   ◊ TASKS DELETED
◊ TASKS CHANGED   ◊ LINKAGE REVISIONS
◊ RESOURCE REVISIONS   ◊ DURATION REVISIONS

❑   EXECUTION PHASE COST TRACKING

❑   EXECUTION PHASE COMPLEXITY TRACKING

❑   EXECUTION PHASE STAFF TRACKING

❑ $TTM_{S/W}$ ACTUAL TRACKING

❑   CONCISE, PERIODIC PROGRESS/PROBLEM REPORT
     TO THE PROJECT/PROGRAM MANAGER

SUGGESTIONS/REQUIREMENTS:

❑   THOROUGH UNDERSTANDING OF THE INFORMATION
      NEEDED BY OTHERS

❑   DEFINED AND COMMUNICATED DOCUMENTATION
      STANDARDS

❑   DEFINED AND STRUCTURED SOFTWARE EVALUATION
      METHODS

❑   ACCESSIBLE TO OTHER TEAM MEMBERS IN NEED OF
      INFORMATION CLARIFICATION

❑   ACCESS TO PRE-MANUFACTURING HARDWARE

❑   TALENTED, MOTIVATED TEAM OF PEOPLE

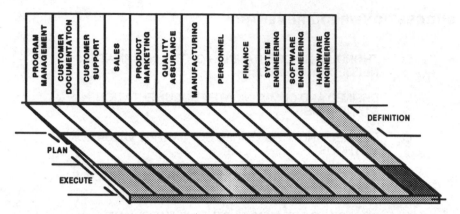

PHASE: EXECUTION   **TEAM**: HARDWARE ENGINEERING

OUTPUT:

❏   PRE-MANUFACTURING HARDWARE AVAILABLE

❏   THOROUGH HARDWARE EVALUATION AND TEST

❏   ACTUAL HARDWARE PERFORMANCE CLOSURE WITH
     PREDICTIONS

❏   DESIGN INFORMATION SUPPLIED TO

     ◊ CUSTOMER DOCUMENTATION   ◊ CUSTOMER SUPPORT
     ◊ PRODUCT MARKETING   ◊ QUALITY ASSURANCE
     ◊ MANUFACTURING   ◊ SYSTEM ENGR
     ◊ SOFTWARE ENGINEERING

❏   PERIODIC REPORT ON NETWORK ACTIVITY TO
     PROJECT/PROGRAM MANAGER

     ◊ TASKS ADDED   ◊ TASKS DELETED
     ◊ TASKS CHANGED   ◊ LINKAGE REVISIONS
     ◊ RESOURCE REVISIONS   ◊ DURATION REVISIONS

❏   EXECUTION PHASE COST TRACKING

❏   EXECUTION PHASE COMPLEXITY TRACKING

❏   EXECUTION PHASE STAFF TRACKING

❏ $TTM_{h/w}$ ACTUAL TRACKING

❏   CONCISE, PERIODIC PROGRESS/PROBLEM REPORT
     TO THE PROJECT/PROGRAM MANAGER

SUGGESTIONS/REQUIREMENTS:

- ❏   THOROUGH UNDERSTANDING OF THE INFORMATION NEEDED BY OTHERS

- ❏   DEFINED AND COMMUNICATED DOCUMENTATION STANDARDS

- ❏   DEFINED AND STRUCTURED EVALUATION EVALUATION METHODS

- ❏   ACCESSIBLE TO OTHER TEAM MEMBERS IN NEED OF INFORMATION CLARIFICATION

- ❏   ACCESS TO PRE-MANUFACTURING SOFTWARE

- ❏   TALENTED, MOTIVATED TEAM OF PEOPLE

CORPORATE BUSINESS DEFINITION EXAMPLE

MONI CORPORATION BUSINESS DEFINITION

Vision

MONI exists to satisfy the personal and home entertainment needs of all people in the world.

Five-Year Objectives

- Determine how to best integrate BCS Records.
- Determine how to best integrate Paramound Pictures.
- Maintain worldwide net profits at greater than 8 percent of sales.
- Grow gross sales at greater than 15 percent per year.
- Grow expenses at no more than 10 percent per year.
- Form alliances with leadership companies with similar business interests.
- Follow the MONI corporate objectives.
- Expand audio market share at greater than 2 percent per year.
- Expand video market share at greater than 3 percent per year.
- Expand systems market share at greater than 5 percent per year.
- Expand personal products market share at greater than 1 percent per year.
- Expand software market share at greater than 7 percent per year.
- Develop at least two new major markets.

Customers and Channels of Distribution

Target customers are located in the United States, Western Europe, and Japan. Our product and service offering will be designed to meet the needs of consumers in each of these primary regions. We will service other regions of the world by selectively modifying products produced for the primary regions. Our product and service offering shall have no age limits. Certain products and services are expected to appeal to children while others cater to the needs of those who are retired from active employment. Our goal is to have MONI products be an integral part of our customers' lives from birth to death. Of course, we must provide products and services adjusted to the economic status of our customers as they age by providing suitably priced and positioned alternatives.

Target customers, male and female, will span a broad age, geographic, and economic range, but we will focus on those with preference and income to obtain products that deliver top performance with high reliability. We want to appeal to those customers that are attracted to our performance, reliability, and quality image while we deemphasize promotion to customers who buy based on low price alone. Likewise, we will not expend a great deal of our effort developing products for those individuals who seek only the highest-performance products available in the market.

Our customers will take delivery of our products from the network of worldwide retailers that we have established to distribute our products. Our retailers will have key strengths in selected market segments that span discounted, high-volume sales through those that supply valuable assistance in product selection and application at relatively low sales volumes. Additionally, we will support a mail/phone order distribution system that appeals to those customers that do not need the added services provided by personal retail outlets. Our retailers will be required to service and support those products that they sell. They may elect to forward to MONI service and support those products that need repair. They may elect to forward complex and specialty repairs to regional service centers that are operated by MONI or to MONI-trained repair specialists that operate in their immediate geographic area.

Competition

Our broad product line has competitors active in many of the markets that we serve. These competitors (RCA, Mitsubishi, Panasonic, Canon, Ricoh, Chinon, Pioneer, Japan Victor Corporation, General Electric, Kyocera, Toshiba, Magnavox, Zenith, Technics, NEC, Sharp, Proton, Sansui, Yamaha, and Phillips) are well known internationally. Most are headquartered in Japan. Firms in the United States (General Electric, RCA, Zenith, and Magnavox) and Europe (Phillips) have formed strong alliances with our Japanese competitors. This list of impressive competitors should keep doing business in our chosen markets exciting. Predicting the next move of these players is very difficult to accomplish, but data are available to show their present market strengths and weaknesses. From this information, and the added information we receive from our retailers and customers, we will construct strategic and financial models that should allow us to gain some insight into the future behavior of these companies.

Necessary Products and Services

Personal Audio Products. Products such as the "Stepman" and its dozens of derivatives. AM/FM headsets, compact disk portable players, compact tape recorders, compact DAT players. Cordless and cellular telephones.

Personal Video Products. Camcorders in VHS, Beta, and 8-mm formats. Personal compact television in the "Lookman" tradition. Additionally, we shall continue the development of still video recorders and players.

Personal System Products. The integration of audio, video, and software into compact personal products. Main emphasis is on "Lookman" technology that contains a VCR and functions with ultracompact video cameras.

Home Audio Products. Home audio components (CD players, tape players/recorders, preamplifiers, amplifiers, speakers) and integrated audio receivers that employ the latest in digital signal processing technology.

Home Video Products. Television receivers and video monitors that employ CRT and projection technology. 8-mm, VHS, and Beta videocassette recorders. Laser disk video players.

Home System Products. Fully integrated systems employing selected home audio and video products that operate under IR or computer control.

Software. Audio and video software that is compatible with personal and home products. Sofware media products such as tape, magnetic disks, and optical disks.

Other. Opportunistic marketing of special products for the computer, automobile, and communication markets. We are becoming very interested in the laptop computer and personal workstation markets. Creative application of these product technologies could help us maintain the realization of our vision in the years to come.

Market Information Summary

Product Line	Market Size ($ B)	Market Growth (% yr)	MONI Growth (% yr)	Market Share (%)	Key Competition
Personal audio	4.7	4.0	5.0	45.0	Panasonic
Personal video	3.3	11.0	17.0	15.0	JVC
Personal systems	0.5	35.0	12.0	9.0	JVC
Home audio	4.1	4.0	22.0	37.0	Panasonic
Home video	18.7	11.0	18.0	44.0	Mitsubishi
Home systems	1.4	42.0	57.0	31.0	Pioneer
Audio software	4.6	11.0	12.0	19.0	Columbia
Computer	9.8	16.0	4.0	11.0	Sun
Auto	1.7	4.0	3.0	7.0	Pioneer

Product Vintage Chart

Produced from the sales forecasts of the Personal Products, Home Products, Software, and Computer groups.

Product-Line Guidance

Personal Audio. The market continues to grow largely due to the continued double-digit demand for portable compact disk players, while other product lines show modest to no growth. MONI will continue to invest in this product line at present rates to maintain market share. Be on the alert to identify low-cost manufacturing sites for headset and tape products. Cordless phone sales are meeting expectations, which suggests we will continue investment in this product line. Results form our entry into cellular phones have been disappointing and will be terminated. Excess profits generated by this product line shall be used to fund other opportunities.

Personal Video. 8-mm camcorders and accessories are the backbone of this product line. "Lookman" compact TV sales growth has been impressive. Early sales results from still video camera sales are encouraging. Market size and growth potential suggest that we will continue to invest in this opportunity above normal levels.

Personal Systems. The present market is small but growing rapidly. Most sales are now going into commercial applications. This is an area where the management team wants to continue investment in new product development.

Home Audio. CD players continue to provide us with double-digit growth. Sales on other products within the line are essentially flat or slightly declining. Look to move the manufacturing of this product line to areas that offer better competitive advantage. Development funding will be slightly reduced over last year.

Home Video. This product line continues to be our largest contributor to cash generation. Sufficient profits are being created to fund other opportunities. Large-screen, high-resolution display sales have been impressive. Similarly, the sales of video receivers that contain audio signal processing have been growing at double-digit rates. Our growth rate continues to outpace that of the market, a trend we want to see continue. Look to make advances in the HDTV arena. Development funding will grow slightly over last year's levels.

Home Systems. The market for fully integrated audio/video systems is growing at an impressive rate. Our strategy to equip medium- and high-performance components with IR and 232 interfaces is being well received by customers. We now own about one-third of the market and have the potential to increase our market share dramatically with little competitive pressure. Though the market seems small at present, our long-term forecasts suggest that it will be a major competitive area in the next two to four years. Investment levels will be increased over last year to accelerate incremental developments and start work on the next generation of home system components.

Audio Software. Our acquisition of BCS Records continues to reward us financially. We will maintain last year's aggressive funding levels to increase our intentional portfolio of recording artists.

Video Software. The acquisition of Paramound Pictures is causing some difficulty. Since under our management, the growth rate has dropped below that of the overall market. Blockbuster new products are needed to turn this situation around. Our attempts to make last year's big hits incrementally better have failed. We will recruit talent from outside MONI that is capable of making major motion picture hits and get Paramound back on track. Funding will continue at the present high level to get this business unit on firm footing.

Computer. We are a dominant force in the high-resolution computer display market and in the low-end, 3.5-inch disk drive market; however, we are losing our market momentum. To regain the initiative, we will focus our attention on

CD-ROM and workstation product development. Profitability in this area will be sacrificed to gain added market penetration. New product development funding will be increased to further secure a firm market position. We continue to struggle with the conflicts created by this product development effort and our corporation's vision.

Auto. Our participation in the auto stereo market is facing stiff competition and heavy losses. Plan to exit from this market. There will be no funding allocated to new product development.

Other. Our research groups continue to investigate opportunistic applications of our technology base. Our research focus will be on computing, communication, display technology, and high-density optical information storage devices. Funding will remain at present levels.

Potential Problem Analysis

Slow growth in the overall audio market is causing some concern. To date, we have been successful in the market by creating innovative products that were well received. New product plans are aggressive and seem to contain a mixture of offerings that will continue our market share growth. We're hopeful that we will continue to be rewarded for our quality image in this market.

Video market growth is slowing. Competition is more intense than ever. We're in a race to get out our first-of-a-kind flat panel display, called "Revolution." Many of our new product development programs are dependent on this technological and manufacturing breakthrough. Its success and our success go together.

The laptop computer and personal workstation market are growing very rapidly. Our advanced research group has developed methods to add significant entertainment value to these emerging technologies. The new product concepts do not fit well within our existing organizational structure. We are considering the possibility of creating a new entity within our Computer group whose entire energy would be on creating a new market.

Division Business Definition

Each division is expected to prepare a one-year, rolling business definition that is updated and reviewed quarterly by the corporate management team.

DIVISION BUSINESS DEFINITION EXAMPLE

MONI DISPLAY DIVISION BUSINESS DEFINITION

Vision

To assist MONI Corporation satisfy the personal and home entertainment needs of all people in the world by providing displays for the MONI family of products.

Five-Year Objectives

- Maintain worldwide net profits at greater than 8 percent of sales.
- Grow gross sales at greater than 15 percent per year.
- Grow expenses at no more than 10 percent per year.
- Form alliances with leadership companies with similar business interests.
- Follow the MONI corporate objectives.
- Grow video market share at greater than 3 percent per year.
- Develop at least one new major market.

Customers and Channels of Distribution

Target end users are located in the United States, Western Europe, and Japan. MONI's product and service offering will be designed to meet the needs of consumers in each of these primary regions. The corporation will service other regions of the world by selectively modifying products produced for the primary regions. MONI's product and service offering shall have no age limits. Certain products and services are expected to appeal to children while others cater to the needs of those who are retired from active employment. Our goal is to have MONI products be an integral part of our customers' lives from birth to death. Of course, we must provide products and services adjusted to the economic status of our customers as they age by providing suitably priced and positioned alternatives.

Target MONI customers, male and female, will span a broad age, geographic, and economic range, but the corporation will focus on those with preference and income to obtain products that deliver top performance with high reliability. MONI wants to appeal to those customers that are attracted to a performance, reliability, and quality image.

The Display Division will reach target MONI customers through products designed, manufactured, and marketed by other MONI divisions. In effect, our direct customers are other MONI product divisions and distribution centers. They will take delivery of our products through the internal MONI parts distribution network.

Competition

Our display product line has competitors active in many of the markets that we service. These competitors (Mitsubishi, Pioneer, Japan Victor Corporation, General Electric, Toshiba, Zenith, NEC, and Sharp) are well known internationally. Most are headquartered in Japan. Firms in the United States (General Electric, Zenith, and Magnavox) and Europe (Phillips) are actively engaged in the development of next-generation and first-of-a-kind display devices to regain market penetration. This list of impressive competitors should keep doing business in our chosen markets exciting. Predicting the next move of these players is very difficult to accomplish, but data are available to show their present market strengths and weaknesses. From this information, and the added information we receive from our retailers and customers, we will construct strategic and financial models that should allow us to gain some insight into the future behavior of these companies.

Necessary Products and Services

We will supply new display products for the following MONI product lines:

Personal video   *Better Lookman:* New product effort will be dedicated to improving resolution and reducing manufacturing costs.

                 Personal display: The new product effort is being directed toward the development of a 9-inch diagonal detachable camcorder color display.

Home video   *3/4 cost:* Steps shall be taken to modify design and manufacturing processes on existing flat panel displays to reduce manufacturing costs by 25 percent.

                 Revolution: A first-of-a-kind development effort will be initiated to bring a revolutionary product to the market.

Computer   *Rainbow:* A new color display for the new series of MONI laptop UNIX workstations.

                 U-Tube (ultimate CRT): A 27-inch diagonal workstation display with the best resolution on the market.

Development and Introduction Plan

Product Line	Qtr 1	Qtr 2	Qtr 3	Qtr 4	Qtr 5	Qtr 6	Qtr 7	Qtr 8
Personal video		Better lookman			Personal display			
Home video				3/4 cost			Revolution	
Computer video			U-tube			Rainbow		

Note: Product definition sheets are to be added to this section of the business definition by the product development teams responsible for each product. The reader of the business definition will then get a full business briefing on each of the accepted new product development projects. Please refer to the example product definition sheet (next document in Appendix) for more insight on the "Revolution" program.

Development Staff and Costs

	Q1	Q2	Q3	Q4	Q5	Q6	Q7	Q8
Personal								
Better Lookman								
R&D people	10	10						
Mfg people	10	10	3					
Mktg people	2	2	2					

	Q1	Q2	Q3	Q4	Q5	Q6	Q7	Q8
Personal								
Personal display								
R&D people	2	2	12	12	12	12		
Mfg people	4	4	14	14	14	14	14	
Mktg people	2	3	4	4	4	4	4	4
Home Video								
3/4 cost								
R&D people	7	7	7	7	7			
Mfg people	10	10	10	10	10			
Mktg people	2	2	2	2	2	2	2	2
Revolution								
R&D people	6	6	21	23	24	25	25	25
Mfg people	12	12	31	35	35	35	35	35
Mktg people	5	5	7	10	10	10	101	10
Computer								
U-Tube								
R&D people	12	12						
Mfg people	20	20						
Mktg people	3	3		3	3	3		
Rainbow								
R&D people	11	11	11	11	11	11		
Mfg people	10	10	10	10	10	10	10	10
Mktg people	4	4	4	4	4	4	4	4
Other								
New projects								
R&D people						8	35	40
Mfg people						10	12	30
Mktg people						5	6	7
R&D Total People	*48*	*48*	*51*	*53*	*54*	*56*	*60*	*65*
Mfg total people	*66*	*66*	*68*	*69*	*69*	*69*	*71*	*75*
Mktg Total People	*18*	*19*	*22*	*23*	*23*	*25*	*26*	*27*
Investment/Qtr ($K)	*3960*	*3990*	*4230*	*4350*	*4380*	*4500*	*4710*	*5010*

Potential Problem Analysis

The major potential problem is the high-risk position the division is assuming on the development of the "Revolution" product. As a hedge against risk, the increase in staffing in Q3 will not take place if Q2 program milestones have not been achieved.

The competitive situation is heating up. Receptor divisions continue to apply price reduction pressure. One division manager has indicated that he may go outside of MONI to obtain a lower-cost supplier. Naturally, we will do all in our power to satisfy the demands of this important customer. Time to market on "3/4 cost" is a top priority for us to satisfy this customer's needs.

"Rainbow" development is highly leveraged by the corporation. We must deliver this high-resolution, low-power color display on time and in large volumes to meet projected market demand for the portable UNIX workstation. If volume manufacturing is not achieved, the resulting slow product buildup will eliminate the possibility of MONI gaining a foothold in the personal computation field.

Recommendations

The display division management team recommends that this business definition be accepted to serve as our new product development guide for the business year and that it be reviewed and updated on a quarterly basis.

Tactical Plan

Each business team is expected to prepare a one-year, rolling tactical plan that is updated and reviewed quarterly by the division management team.

EXAMPLE NEW PRODUCT DEFINITION SHEET (NPDS)

M NEW PRODUCT DEFINITION MONI DISPLAY DIVISION	MONI 2001 FLAT COLOR DISPLAY (REVOLUTION)

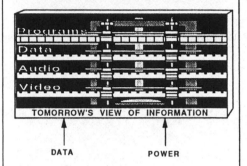

TOMORROW'S VIEW OF INFORMATION

DATA   POWER

INTERNAL STRATEGIC PURPOSE:

A next generation development effort to keep MONI technologically ahead of the competition and retain market dominance. A display technology for all future home and computer products.

EXTERNAL STRATEGIC PURPOSE:

A new video presentation screen that provides unsurpassed information display capability.

PRODUCT DESCRIPTION:

A large, thin, high resolution display.

CUSTOMERS:

❏ MONI Home Products Group

❏ MONI Computer Group

COMPETITIVE POSITION:

❏ Small United States Firms include:

* Photonics Technology
* Ovonic Imaging Systems
* Magnascreen Corporation

❏ Large Firms include:

* Zenith
* Phillips
* NEC
* Toshiba

Competitors do not have MONI's name or advanced distribution system, but seem to be well staffed and suitably funded. There is a worldwide race taking place to develop the flat screen technology. Recruiting technical talent from the competition is desirable.

KEY FEATURES:

❏ 30 x 18 x 3 Minimum Size

❏ Color

❏ HDTV Quality Resolution

❏ Computer Graphics Compatible

SPECIFICATIONS:

❏ 900 X 1500 Pixels

❏ 16 Bit Resolution

❏ 95 Watts Power Consumption

❏ Maximum Sizes Up To 150 X 90 X 5 Inches

RESOURCE REQUIREMENTS:

	PERSONS	PERSON MONTHS	$M
R&D	23	155	$1.9
MFG	35	230	$2.1
MKTG	10	67	$.5
TOTAL	68	452	$4.5

SALES:

SALES PRICE:	$130
MATURE SALES:	120K Units/Month
MARKET WINDOW:	1 Year +/- 50%
EFFECT ON OTHER PRODUCTS:	Major

FACTORY COST:	$95
IRR:	23%

SCHEDULE:

Definition Phase Complete	Month 9
Planning Phase Complete	Month 11
Execution Phase Complete	Month 23
Product Shipment	Month 26

RISKS AND CRITICAL ISSUES:

❏ Concurrent development of design and manufacturing.
❏ Product introduction > 1 year after competition.
❏ Exotic material supply remains adequate.
❏ Consumer demand for this type of product.

EXAMPLE CORPORATE OBJECTIVES

OUR CUSTOMERS

To provide products and services of the greatest possible value to our customers, thereby gaining and holding their respect and loyalty.

OUR PRODUCTS

To let our growth be limited by our ability to develop and produce products of lasting value that satisfy real customer needs.

OUR PEOPLE

To attract and develop only the best individuals for critical positions within the company. To enable these people to share in the company's success, which they make possible.

OUR PROCESSES

To develop and advance only key processes that ensure long-term competitiveness of the company.

MANAGEMENT

To foster initiative and inspire freedom of imagination by allowing the individual great freedom of action in attaining well-defined corporate and performance objectives; to provide an atmosphere of creativity and fun.

CITIZENSHIP

To honor our obligations to society by being an economic, intellectual, and social asset to the community in which we operate.

PROFIT

To achieve sufficient profit to finance our company growth and to provide the resources we need to achieve our other objectives.

TIME-TO-MARKET COST MODEL

CALCULATING THE COST OF LATENESS

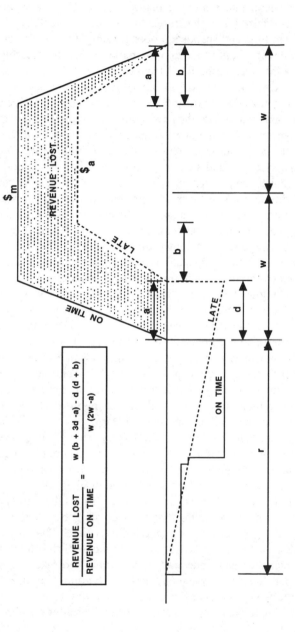

$$\frac{\text{REVENUE LOST}}{\text{REVENUE ON TIME}} = \frac{w(b + 3d - a) - d(d + b)}{w(2w - a)}$$

The above model was formulated to place new product development financial issues into perspective. Expenses in the early stages of new product development are relatively small. It is likely that an individual or small team will generate an idea that will then become staffed over a period of time. This early portion of the investment triangle is referred to as the "research and development" segment. This group grows in size over time and spends a proportionately greater sum of money until the point is reached where staff is added from marketing, manufacturing, and support groups. At this point in time, the R&D team is likely to reach its maximum size, while the total staff continues to grow as other functions join the new product development effort. With the addition of staff comes the corresponding increase in the production of output associated with the new product development process. Each function will begin contributing to the new product process. Costs will increase beyond just the direct labor and overhead rates associated with the people doing the work. Tools, materials, systems, and support groups are added as the new product development process makes the product ready for its market introduction. In general, it is easy to see the mechanisms at work that produce the triangular shape of the investment model. Investment growth is largely attributed to an increase in staff over the length of the project and the resultant expenses that each staff member can generate. The "ramp" new product development staffing profile is the most common.

In contrast to the ramp staffing profile, consider the profile recommended in this text. Both structures contain the same area under the curve and have the same maximum staffing level; yet, the "on time" profile delivers product to market considerably earlier than the ramp profile. This on-time profile suggests that approximately 50 percent of the development effort goes into getting the product's definition "right" using about 20 percent of the full staff. Once the definition work is done, a brief planning period begins that should consume three to six weeks of calendar time to plan the execution phase thoroughly. With acceptable plans in hand, the project is then rapidly staffed to full cross-functional levels and executed. The execution phase then draws upon a full staff using the best available tools and methods, and, working as a unified and focused team, gets the fully defined product to market. The total "on time" investment is proportional to the area under the "on time" curve. Likewise, the total "late" investment is proportional to the area under the "late" curve. Subtracting the "on time" area form the "late" area yields the differences in investment made between the two profiles.

Rarely is it possible to spend less money to get the same product to market. This is just another way of saying that so much effort has to go in to doing the same job regardless of how that effort is applied over time. The only alternative available is to decide how to staff the effort to do the work. If the effort is poorly defined, marginally planned, and understaffed, it is going to take longer than a fully staffed project that follows a systematic process. Examination of proposed, or sometimes actual, staffing profiles of the project is the key to understanding if the project is likely to have a late entry to market. "Ramp" profiles are almost always late. "Step" profiles are highly likely to be on time. To convince yourself, take the rapid development projects done by your organization and reconstruct their full cross-functional staffing profiles. Compare these profiles to those of projects that had delayed market entry. You'll soon develop the sensitivity needed to staff projects effectively to achieve market-entry requirements.

"Ramp" profiles, besides taking longer, often consume more development funds. This happens whenever work is redefined and redirected—characteristic of projects that are poorly defined and planned. This waste of person-months costs time and money. Therefore, it is appropriate to scale "ramp" profile total expenses up by a factor over "step" profile total expenses. The scaling factor is a figure that is likely to be in the range of 0.1 to 1; that is, a "ramp" staffed effort is likely to cost from 10 to 100 percent more than a "step" staffed project as it enters the market late! Determination of the appropriate scaling factor for the project being analyzed is usually a very simple exercise for an organization that understands its product development cost structure.

After following the above procedure, it is not unusual to find that it costs less to get products to market on time than it does to get them there late. At the same time, the "step" profile gives the organization the option to cancel a project up to its 50-percent point in time and sacrifice less than 20 percent of its development investment. Thus, with just a small amount of work, you can determine the financial impact that being on time has on the investment side of the new product development equation.

Financial benefits for on time entry continue as the product moves into its revenue phase. The equation associated with the figure makes it easy to determine the ratio of lost revenue to on-time revenue. The model has been structured so that new product development team participants can work with it easily to reach consensus on model parameters. A cross-functional team merely has to estimate values for the market window w, the time to rise and fall from mature on time sales a, and the time to rise and fall from mature late market-entry sales b. Merely insert the numerical values for these figures into the equation and compute the ratio for various values of d. Once you do, you'll begin to understand why on-time market entry is so important!

For example, consider the possibility of entering the notebook computer market with the latest available chip technology from Intel. Based on the turnover rate of the technology content of the product and some knowledge about the market, it's possible to determine that $12 < w < 24$. For analysis purposes, it was also easy to let $a = b = 6$. Under worst-case conditions ($w = 12$), being one month late sets the ratio at 0.16. Two months late puts it at 0.33. For the best case ($w = 24$), one month late sets the ratio at 0.06 and two months at 0.13. The model suggests to this team of developers, no matter what their assumptions are about w, their best strategy is to be on time. If this product idea has a total revenue potential of $100 million, being a few months late can cost the organization from $6 million to $33 million, depending on conditions. Should this team need to spend added funds to be on time, it's best for them to do so!

The model scales the revenue axis of the late sales profile to decline by the relationship of $(1 - d/w)$. This relationship set (revenue lost/revenue on time) to equal 1 ($a = b$) when $d = 0$. By the time $d = w$, the ratio equals zero. If this penalty for being late seems too severe, I suggest using a market window scaling relationship that is more suitable to your needs. To make being late look best, don't scale the revenue axis of the late sales profile at all. In this way, the ratio will increase proportional to $d/2w$. Under this condition, being one month late, with $w = 24$, results in a late to on-time ratio of 0.02. So, even with trying to make the model make the situation look better, it still looks bad.

In general, you can spend a fair amount of money to be on time and, if successful, gain considerable financial benefit. Combining investment and revenue strategies to yield timely delivery of product to market is certain to improve the financial performance of the business.

SOFTWARE EVALUATION

VIEWPOINT, VER 4.1

This is an evaluation of the project management program ViewPoint, Ver. 4.1, by Computer Aided Management, Inc. (CAM). ViewPoint was evaluated against the tool selection criteria established in Chapter 5. Use it as a guide and a reference when performing an evaluation of other software alternatives.

Star System of Criterion Evaluation

☆☆☆☆☆ = Best   ☆ = Worse

Tool Selection Criteria

1. *Supports Hierachical Planning Techniques.*   ☆☆☆☆☆
   ViewPoint was designed around the hierarchical breakdown of networks. Using ViewPoint makes hierarchical planning intuitive.

2. *Functions Effectively on >1000 Tasks.*   ☆☆☆☆☆
   Viewpoint supports 32,000 activities.

3. *Usable by Horizontal and Vertical Decision Makers.*   ☆☆☆☆
   The user interface for ViewPoint is a character-graphics screen, horizontally time-scaled. The user gets immediate visual feedback when defining and linking activities, greatly enhancing user confidence. Some users would like to see a "PERT" view, where each activity is represented in the network by a fixed-size box. ViewPoint does not support an on-screen "PERT" view, but the graphics package outputs "PERTS," Gantt, and networks.

4. *Usable by Support Personnel.*   ☆☆☆☆
   Hypersensitivity to mouse movements when in table entry mode, and a medium-sized learning curve for data-entry personnel are the only detractors in this area. The graphical user interface closely emulates the appearance of the network layout sheets within PLASTK, so data-entry errors are easily noticed and corrected. An on-line loop detec-

tion function prevents loop errors when establishing precedence relationships.

A data-entry benchmark project of 100 activities took two hours to enter. This is the standard by which to compare other project management software.

5. *Supports Tracking by Storing a Reference.*   ☆☆☆☆☆
   Viewpoint stores a baseline reference, as well as the current plan, and actuals. The "baseline" may be set at any time during the course of the project.

6. *Supports Merging of Projects and Subprojects.*   ☆☆☆☆☆
   Any other project may be inserted into a branch of the hierarchy; each hierarchical branch may become a separate project. This enables resource leveling across multiple projects.

7. *Does Calculations Fast.*   ☆☆☆☆
   In "Auto Calc" mode, where each modification causes a recalculation, the data-entry process starts slowing down after a few hundred activities are entered. Putting the system into "Manual Calc" mode, or "Local" mode, where only the current branch of the hierarchy is calculated, keeps the system lively.

8. *Supplies the Ability to Export Data.*   ☆☆☆☆☆
   The export/import feature is well thought out. The user can export any subset of data, based on dates, status, and so on, in a variety of formats, including ASCII, Lotus, and dBase. Once set up, an export/import definition can be saved in a library for later reuse.

9. *Attention Paid to Laser Printer Support.*   ☆☆
   Supports HP LaserJet, but not PostScript printers.

10. Supports Graphical Input of Networks.   ☆☆☆☆☆
    Very nicely.

11. *Support for Table Entry of Task Data.*   ☆☆☆☆
    Yes. When in table-entry mode, however, it is better to use the cursor keys to navigate, and leave the mouse alone.

12. *Calculates Task Durations from Work Quantity and Resource Commitment.*   ☆☆☆☆☆
    Yes. Additionally, it allows the user to vary the commitment over time.

13. *Performs Resource Leveling and Constraining.*   ☆☆☆☆
    Constraining is done graphically on a resource usage histogram. The user draws a line depicting the maximum availability (dropping the line to zero during vacations, etc.), then pushes the Go button. Leveling does what it can by eating up float only: no changes are made to affect the project finish date. Constraining is more drastic: activities and project completion date will be pushed to accommodate the constraint.

BIBLIOGRAPHY

Amabile, Theresa M., *Growing Up Creative,* Crown, New York, 1989.

Brooks, Frederick, *The Mythical Man-Month.* Addison-Wesley, Reading, MA, 1975.

Burgelmann, Robert A., and Sayles, Leonard R., *Inside Corporate Innovation: Strategy, Structure, and Management Skills.* Free Press, New York, 1986.

Chen, C. T., *Introduction to Linear System Theory,* Holt, Rinehart, & inston, New York, 1970.

DeMarco, Tom, and Timothy Lister, *Peopleware,* Dorset House, New York, 1987.

Eisenhardt, Kathleen M., "Speed and Strategic Choice: How Managers Accelerate Decision Making, *California Management Review,* Vol. 32, No. 3, Spring 1990.

Evans, Michael W., *Principles of Productive Software Management,* Wiley, New York, 1983.

Fortune Magazine, February 13, 1989.

Heller, Robert, *The Decision Makers:* The Men and the Million-Dollar Moves: Behind Todays Great Corporate Success Stories, Truman Tally Books, New York, 1989.

Hewlett-Packare Annual Report, New York, 1988.

Hughes Aircraft Company, *R&D Productivity,* Culver City, CA, Hughes Aircraft Company, 1978.

Jones, Capers, *Programming Productivity: Issues for the Eighties,* IEEE Catalog No. EH0186–7, IEEE, New York, 1981.

Kidder, Tracy, *The Soul of a New Machine.* Little, Brown, Boston, 1981.

Kotler, Philip, *Marketing Management: Analysis, Planning, and Control.* Prentice-Hall, Englewood Cliffs, NJ, 1984.

Larson, Gary, *The PreHistory of The Far Side.* Andrews & McMeel, Fairway, KS, 1989.

Levine, Harvey, *Project Management Using Microcomputers,* McGraw-Hill, New York, 1986.

Randolph, Alan, and Barry Z. Pozner, *Effective Project Planning and Management: Getting the Job Done.* Prentice-Hall, Englewood Cliffs, NJ, 1988.

Smith, Preston G., and Donald G. Reinertsen, *Developing Products in Half the Time,* Van Nostrand Reinhold, New York, 1991.

Weinberg, Gerald, *An Introduction to General Systems Thinking,* Wiley, New York, 1975.

White, John A., Agee, Marvin H., Cuse, Kenneth E., *Principles of Engineering Economic Analysis,* Wiley, New York.

INDEX